EPA/600/R-13/354
December 2013
www.epa.gov/ord

A Review of Health Impact Assessments in the U.S.: Current State-of-Science, Best Practices, and Areas for Improvement

Justicia Rhodus[1], Florence Fulk[2], Bradley Autrey[2], Shannon O'Shea[3], Annette Roth[2]

[1] CSS-Dynamac
c/o U.S. Environmental Protection Agency
Cincinnati, OH 45268

[2] U.S. Environmental Protection Agency
National Exposure Research Laboratory
Cincinnati, OH 45268

[3] Contractor, U.S. Environmental Protection Agency
National Exposure Research Laboratory
Research Triangle Park, NC 27709

National Exposure Research Laboratory
Office of Research and Development
U.S. Environmental Protection Agency
Cincinnati, OH 45268

Notice

The U.S. Environmental Protection Agency through its Office of Research and Development funded and managed the research described here under contract EP-D-11-073 to Dynamac Corporation. It has been subjected to the Agency's external and administrative reviews and has been approved for publication as an EPA document.

Author Contributions and Acknowledgements

J. Rhodus (HIA Review Lead) – developed and managed the HIA Review process (including review framework and database), reviewed HIAs, implemented the QA Review and corrective actions, compiled and managed the final review database, performed final quality check on review database, synthesized the results of the HIA Review, and authored this synthesis report.

F. Fulk (Principal Investigator) – managed implementation of the HIA Review in the context of other related Agency research, provided input during conceptualization of the HIA Review process, reviewed HIAs, and reviewed this synthesis report.

Brad Autrey (HIA Review Team) – provided input during conceptualization of the HIA Review process, reviewed HIAs, and reviewed this synthesis report.

Shannon O'Shea (HIA Review Team) – reviewed HIAs and reviewed this synthesis report.

Annette Roth (HIA Review Team) – reviewed HIAs and reviewed this synthesis report.

The authors would also like to thank Ellen D'Amico (CSS-Dynamac) for GIS support and Bruce Mintz (EPA-NERL), Aaron Wernham (Health Impact Project), Celia Harris (Human Impact Partners), and Steve White (Oregon Public Health Institute) for providing critical review of this synthesis report.

Executive Summary

A systematic review of health impact assessments (HIAs) from the U.S. was conducted to obtain a clear picture of how HIAs are being implemented nationally and to identify potential areas for improving the HIA community of practice. The review was focused on HIAs from the four sectors that the U.S. Environmental Protection Agency's (EPA's) Sustainable and Healthy Communities Research Program has identified as target areas for empowering communities to move toward more sustainable states. These four sectors are Transportation, Housing/Buildings/ Infrastructure, Land Use, and Waste Management/Site Revitalization.

The review systematically documented organizations involved in conducting the HIAs; funding sources; the types of community-level decisions being made; data, tools, and models used; self-identified data needs; methods of stakeholder engagement; pathways and endpoints; characterization of impacts; decision-making outcomes and recommendations; monitoring and follow-up measures; prioritization methods employed; HIA defensibility and effectiveness; attainment of the *Minimum Elements of HIA*; areas for improvement; and identification of best practices. The results of the HIA reviews were synthesized to identify the state of the HIA practice in the U.S., best practices in HIAs, and areas in the overall HIA process that could benefit from enhanced guidance, strategies, and methods for conducting community-based risk assessments and HIAs.

While HIAs have helped to raise awareness and bring health into decisions outside traditional health-related fields, the effectiveness of HIAs in bringing health-related changes to pending decisions in the U.S. varies greatly. The review found that there are considerable disparities in the quality and rigor of HIAs being conducted. This, combined with the lack of monitoring, health impact management, and other follow-up in the HIAs could be limiting the overall utilization and effectiveness of this tool in the U.S. However, a number of best practices were identified in the review, which (if implemented) could help advance the HIA field of practice, reduce disparities in the quality and rigor of HIA, and improve the overall effectiveness of the tool.

HIA is a relatively new and rapidly emerging field in the U.S. Understanding the current state and applicability of HIAs in the U.S., as well as best practices and areas for improvement, will help to advance the HIA community of practice in the U.S., improve the quality of assessments upon which stakeholder and policy decisions are based, and promote healthy and sustainable communities.

Foreword

The U.S. Environmental Protection Agency's (EPA's) National Exposure Research Laboratory (NERL) conducts human and ecological exposure research that provides the tools necessary for EPA to carry out its mission. Critical to the success of the NERL research program is communication and utility of its research to influence and impact decisions aimed towards protecting human health and the environment.

Health impact assessments (HIAs) are becoming a more commonly used tool in the U.S. for incorporating health considerations into the decision-making process of plans, projects, programs, and policies. The review of HIAs in the Transportation, Housing/Buildings/ Infrastructure, Land Use, and Waste Management/Site Revitalization sectors was conducted to inventory the types of community-level decisions being made and to assess the data, tools, and models currently used in HIAs in these four sectors. This information will aid in promoting existing EPA tools, methods, and models that can support HIAs, identifying potential research focus areas to support and improve the HIA community of practice, and discovering what the ecological assessment community of practice could draw from HIAs and vice versa in order to promote healthy and sustainable communities.

Table of Contents

Tables

Figures

Introduction

A systematic review of health impact assessments (HIAs) from the United States (U.S.) was conducted to obtain a clear picture of how HIAs are being implemented nationally and to identify potential areas for improving the HIA community of practice. The review was focused on HIAs from four sectors that the U.S. Environmental Protection Agency's (EPA's) Sustainable and Healthy Communities Research Program (SHCRP) has identified as targets for empowering communities to move toward more sustainable states: Transportation, Housing/Buildings/Infrastructure, Land Use, and Waste Management/Site Revitalization.

A review framework was developed to systematically document:

- organizations involved in conducting the HIAs;
- funding sources;
- the types of community-level decisions being made;
- data, tools, and models used;
- self-identified data needs;
- methods of stakeholder engagement;
- pathways and endpoints;
- characterization of impacts;
- decision-making outcomes and recommendations;
- monitoring and follow-up measures;
- prioritization methods employed;
- HIA defensibility and effectiveness;
- attainment of the *Minimum Elements of HIA*;
- areas for improvement; and
- identification of best practices.

The results of the systematic HIA reviews were recorded in a Microsoft Access database and these results were synthesized to identify the state of the HIA practice in the U.S., best practices in HIAs, and potential areas for improvement.

This report will provide background information on the HIA community of practice, sectors chosen for examination, and methodology employed in the HIA Review, as well as a synthesis of the results of the review and a discussion of what those results mean for the HIA community of practice. In an effort to improve the quality of assessments upon which stakeholder and policy decisions are based and promote healthy and sustainable communities, possible steps to advance the HIA community of practice in the U.S. (e.g., the use of existing tools, methods, and models) will also be identified.

1

Health Impact Assessment: The Tool

HIA Defined

The National Research Council (2011) defines HIA as:

> …a systematic process that uses an array of data sources and analytic methods and considers input from stakeholders to determine the potential effects of a proposed policy, plan, program, or project on health of a population and the distribution of those effects within the population. HIA provides recommendations on monitoring and managing those effects.

This definition is an adaption of the definition developed by the International Association of Impact Assessment (IAIA; Quigley et al. 2006) and is based on a review of HIA definitions, practices, guidance, and peer-reviewed literature.

HIA Steps

There are typically six steps in conducting an HIA (North American HIA Practice Standards Working Group 2010; Bhatia 2011; National Research Council 2011; Human Impact Partners 2011, 2012).

1. **Screening** – Determine whether an HIA is needed and the value added.
2. **Scoping** – Identify which health effects to consider and set the HIA parameters.
3. **Assessment** – Collect qualitative and quantitative information to create a profile of existing health conditions, and identify, evaluate, and prioritize the potential health impacts of the decision.
4. **Recommendations** – Identify alternatives to the decision and/or strategies for promoting the positive health impacts and/or mitigating the adverse health impacts.
5. **Reporting** – Write a final report and communicate the results of the HIA to decision-makers and other stakeholders for implementation/action.
6. **Monitoring and Evaluation** – Evaluate the processes involved in the HIA, the impact of the HIA on the decision-making process, and the impacts of the decision on health.

History of HIA

The HIA community of practice has long been established in Europe, but is a rather young and emerging field in the U.S. While the 1969 National Environmental Policy Act (NEPA) was not an impetus for HIA, it recognized early on the need to consider the health consequences of decision-making. NEPA requires the U.S. government to give consideration to environmental and human health effects prior to undertaking any major federal action (e.g., proposals to adopt rules and regulations, formal plans that direct future actions, programs, and specific projects) that significantly affects the quality of the human environment. The NEPA requirement has helped to generate a great number of Environmental Impact Assessments (EIAs), but historically health impacts have not been adequately addressed in these EIAs (Bhatia and Wernham 2008; see *HIA and Environmental Impacts* for a more in-depth discussion of health considerations in EIA).

More recently, health practitioners, scientists, and decision-makers have recognized that the health of an individual is determined not only by the health care they receive, but also by the natural, social, physical, economic, and political environment in which they live and work. As such, decisions outside of traditional health-related fields can and often do, in fact, influence an individual's health. The recognition that human health can be directly and indirectly impacted by these various factors points to the need for health considerations in decision-making (National Research Council 2011).

In 1986, the World Health Organization (WHO) held the first international conference on health promotion, *The Ottawa Charter for Health Promotion*, and in December 1999 issued the *Gothenburg Consensus Paper* (WHO 1999), which outlined the main concepts and suggested approaches to conducting HIAs.

The first HIA in the U.S. was conducted in San Francisco in 1999 (Bhatia and Katz 2001) and by 2007, twenty-seven (27) HIAs had been completed nationwide (Danneberg et al. 2008). According to the Health Impact Project (2013), the number of HIAs conducted in the U.S. has increased more than eight-fold in the past five and a half years, from 27 HIAs in 2007 to over 225 in early 2013. Health impact assessments have not only became more prominent in the U.S. since 2000 (National Research Council 2011; Health Impact Project 2013), but also worldwide as the World Bank began requiring HIAs for large projects (World Bank Group 2006) and major industries such as oil, gas, and mining (IPIECA/OGP 2000; ICMM 2010) began incorporating HIAs into best business practices. A number of organizations in the U.S have begun promoting the use of HIAs as well, including the Centers for Disease Control and Prevention (CDC), Association of State and Tribal Health Officials (ASTHO), National Association of County and City Health Officials (NACCHO), National Network of Public Health Institutes (NNPHI), the National Research Council (NRC), and others.

Health impact assessment has been promoted worldwide as a tool for protecting and promoting public health because of its applicability in a broad range of decision-making arenas, consideration of beneficial and adverse health consequences, stakeholder and community engagement, and potential to advance health equity (National Research Council 2011).

HIA Standards and Guidelines for the Americas

In 2008, HIA practitioners from Habitat Health Impact Consulting (Canada), the San Francisco Department of Public Health, Human Impact Partners, and the Alaska Native Tribal Health Consortium organized and held the first North American Conference on Health Impact

Assessment. A set of practice guidelines for HIAs was developed by a working group established at that conference, and in 2010, an updated version of those guidelines was issued. The *Minimum Elements and Practice Standards for Health Impact Assessment* identifies essential (i.e., minimum) elements that constitute an HIA and benchmarks (i.e., practice standards) for how best to conduct an HIA (North American HIA Practice Standards Working Group 2010).

In addition to the *Minimum Elements and Practice Standards* and a plethora of existing international guidance, a number of other guides have been developed to inform and direct the HIA practice in the U.S. In recent years, the U.S. has also seen HIA courses popping up in graduate school curriculums, the emergence of HIA technical assistance and training providers, and the dissemination of tools and templates to be used in HIA. Below is a small selection of the guides available to inform HIAs in the U.S.

- *Improving Health in the United States: The Role of Health Impact Assessment* (National Research Council 2011)

- *A Health Impact Assessment Toolkit: A Handbook to Conducting HIA, 3rd edition* (Human Impact Partners 2011)

- *Health Impact Assessment: A Guide for Practice* (Bhatia 2011)

- *HIA Summary Guides* (Human Impact Partners 2012)

- *Rapid HIA Toolkit* (Design for Health 2008)

- *Technical Guidance for Health Impact Assessment (HIA) in Alaska* (Alaska Department of Health and Human Services 2011)

Health Impact Assessment Review

Preliminary Literature Search

A preliminary literature search was performed in early 2012 via the internet to identify HIAs conducted in the U.S. in each of the four chosen sectors: Transportation, Housing/Buildings/ Infrastructure, Land Use, and Waste Management/Site Revitalization. Five primary sources were used in the search (Figure 1). Note that although Figure 1 denotes the WHO website, this was not a fruitful source for our purposes, since most of the HIAs it lists were performed in Europe and our scope was limited to HIAs in the U.S.

Figure 1. Sources used in our literature search to identify HIAs for review.

With the exception of Human Impact Partners, each of these sources assigns a sector designation for the HIAs included on their website. While searching these sources, it was discovered that the terminology for several of the sectors (defined by EPA 2011 and) targeted for this review did not align with the sector terminology used by the organizations. As shown in Figure 2, the sector terminology used among the organizations themselves was also inconsistent. Because of these inconsistencies, searching the sources by sector was not an option. For a more detailed discussion of sector terminology in HIA, see *Consistency in HIA Terminology*.

Health Impact Assessment (HIA) Title	Location	Health Impact Project	HIA Gateway	UCLA HIA Clearinghouse
6th Avenue East Duluth HIA	Duluth, MN	Built Environment		
A Health Impact Assessment of Mixed Use Redevelopment Nodes and Corridors in Lincoln, Nebraska.	Lincoln, NE		Planning/Transport	
Arctic Outer Continental Shelf Oil and Gas Multiple Lease Sale Environmental Impact Statement	Alaska	Natural Resources and Energy	Industry	
Atlanta Beltline	Atlanta, GA	Built Environment	Regeneration	Land Use Planning
Baltimore Comprehensive Zoning Code Rewrite	Baltimore, MD	Built Environment		Land Use Planning
Baltimore Red Line Transit Project	Baltimore, MD	Transportation		Land Use Planning/Transportation
Battlement Mesa	Battlement Mesa, CO	Natural Resources and Energy		Mining, Petro, Other Extractive Industry
Benton Accessory Dwelling Units	Benton County, OR	Housing		
Bernal Heights Preschool	San Francisco, CA	Built Environment		
Buford Highway and NE Plaza Redevelopment	Atlanta, GA	Transportation	Transport	Transportation
Child Health Impact Assessment of Energy Costs and the Low Income Home Energy Assistance Program (LIHEAP)	Massachusetts	Natural Resources and Energy		Utilities
Child Health Impact Assessment of the Massachusetts Rental Voucher Program	Massachusetts	Housing	Housing/Inequalities	Housing

Figure 2. Inconsistencies in sector terminology.

The title and description, if provided, of the HIAs in each of the sources identified in Figure 1 were examined to determine if the HIAs seemed to fall within one of the four sectors, as defined in the Sustainable and Health Communities Research Program (SHCRP) Draft Research Framework (EPA 2011). For a description of the four SHCRP sectors, see *Appendix A* (*pgs A-9 through A-14*). If the HIA seemed to fall within one of the four sectors, the HIA was included in the preliminary literature search results. Using this approach, 78 completed, 2 draft, 66 in-progress, and 3 HIAs of undetermined status were identified in the four sectors. Based on those preliminary results, it was estimated that there could be between 80 and 150 HIAs to review.

Pilot Review

A pilot review of four HIAs (Figure 3), one from each sector, was completed in early spring 2012, following the preliminary literature search. Through this process, the information to be recorded in the HIA Review was refined and the review framework and database were developed.

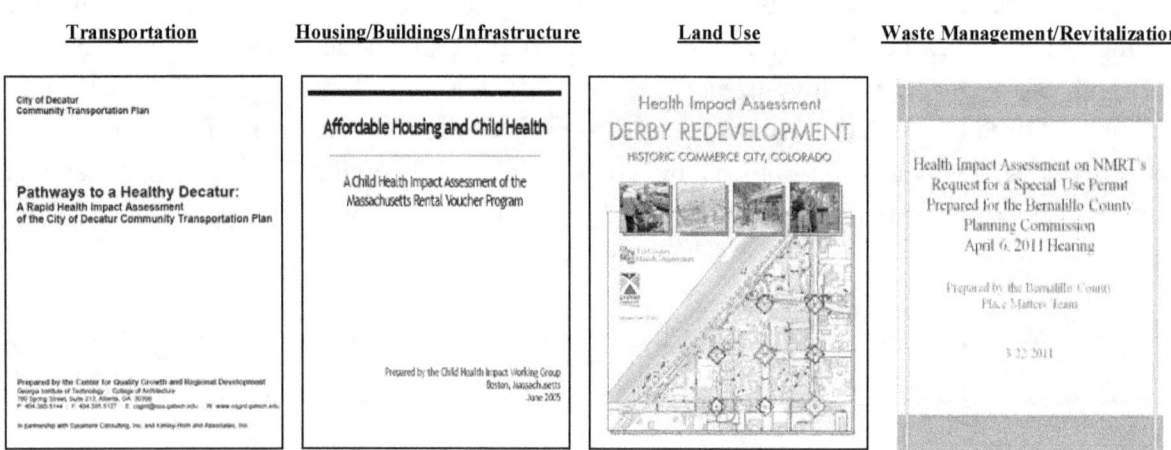

Figure 3. Pilot review HIAs.

Review Framework and Database

Based on the review of the four HIAs in the pilot review, the goals of the project, and the designs of existing HIA databases, a list of proposed data entry fields were developed for inclusion in the review framework.

The *Minimum Elements and Practice Standards for Health Impact Assessment* (North American HIA Practice Standards Working Group 2010) and a number of other ancillary sources were chosen from the broad body of existing HIA guidance to guide the HIA Review. The *Practice Standards* identified in the 2010 guidance document, like much of the available HIA guidelines, do not represent "rigid requirements (for implementation of HIA), but rather reflect an ideal of practice" (from National Research Council 2011). The flexibility offered in this and other HIA guidance acknowledges the diversity of conditions and settings in which HIAs are being performed in the U.S. (i.e., the broad range of decisions being considered and the resource, capacity, and expertise constraints present in implementation). However, the North American HIA Practice Standards Working Group (2010) also acknowledges that there are certain *Minimum Elements* that HIAs must include to distinguish them from other forms of assessment. While there may not be consensus within the field of practice that these elements must be achieved in every HIA, the *Minimum Elements* do provide an ideal of practice, and as such, provided a benchmark against which the HIAs in this review were evaluated.

Following review by the project team, the list of fields in the review framework was finalized and a description of the data to be entered into each field was added to create consistency among the reviews (Table 1).

Table 1. HIA Review Framework. Gray highlights indicate specific terms and/or formatting to be used in data entry

Database Field	Description/Examples
ID	(Automatically generated in Access database)
Title	Full title of HIA Report
Year	Year of publication
Location	Where HIA was conducted– city, county, state, etc. (as applicable)
Decision-making Level	Local, county, state, federal
Organization(s) Involved	Organizations involved in conducting/publishing/sponsoring the HIA
Organization Type	Educational institution, Government agency, Non-profit, Other, Undetermined
Contact	Name and contact info for HIA point-of-contact (if available) in format: name, email / Undetermined
Organization/HIA Website	Identify website dedicated to or highlighting the HIA (if applicable) / N/A
Funding	Identify financial sponsors (if named) / Undetermined
Status	Complete, In progress, etc.
Sector(s)	Transportation, Housing/buildings/infrastructure, Land use, Waste management/site revitalization (as defined by SHCRP)
HIA Type	Mandated (by what/whom), decision support, advocacy, community-led[1]
HIA Rigor	Desk-based, rapid, intermediate, comprehensive[2]
Scope/Summary	Question/problem faced, proposed policy/plan examined
Source of Evidence	Literature review, community consultation, policy review, special collection (interviews, surveys, focus groups, risk assessment, demographics analysis, modeling, etc.)[2]
Data Types	Models, literature (published, peer-reviewed, grey lit, government documents, policy), websites, data
Major Data Sources	Specific models, agency (e.g., CDC, HUD, Census Bureau*) or community data, bibliographic resources (Medline, Pub Med, Web of Science, Science Direct, etc.), databases, websites, internet gateways/search engines (e.g., Google), surveys, focus groups/forums, entities interviewed/consulted (e.g., stakeholders, technical experts), etc. *Note: Note the type, year, and geographic scale of census (and other) data used.
Local Data Available or Obtained?	(If yes) Identify data / No

[1] Harris-Roxas and Harris (2011)
[2] Harris, Harris-Roxas, Harris, and Kemp (2007)

Table 1. Continued

Database Field	Description/Examples
Additional Data Needed (Self-Identified)	(If yes) Identify data / No
Stakeholder/Community Involvement?	(If yes) Identify stakeholder groups* / No *Note: Per the Quality Assurance Review, avenue of involvement to be noted as well.
Impacts/Endpoints	Health (physical, mental, developmental), environmental/ecosystem, behavioral, economic, infrastructure, services, demographic, other
Health Endpoints	Identify health endpoints examined in HIA
Pathway of Impact	Air quality, community/household economics, education, exposure to hazards, healthcare access/insurance, housing, infectious disease, land use, lifestyle, mental health, mobility/access to services, noise pollution, nutrition, parks and recreation, physical activity, public health services, safety (personal, traffic, etc.) and security, social capital, soil quality, water quality, etc.
Characterization of Impact* *Note: Originally labeled Quantification of Impact, but characterization of impacts is both qualitative and quantitative.	Direction (positive, negative, unclear, no effect), permanence, magnitude, likelihood (definite, probable, speculative, unlikely, uncertain), distribution/equity,[2,3] etc.
Decision-making Outcome	Describe the general outcome of the HIA, including recommendations, mitigations, etc.
HIA Report	(Attach HIA Report)
Prioritization Methods* *Note: Originally labeled Impact Prioritization, but methods for prioritizing impacts and recommendations both recorded.	What methods/data were used to prioritize the impacts to be considered [and the recommendations to be developed]?
Defensibility/Process Evaluation	Describe the quality of evidence and methodology; identify assumptions, limitations, barriers; etc.
Effectiveness of HIA	Impact evaluation (direct, general, opportunistic, none[4]), health outcome evaluation (predictive accuracy, health impacts) / Undetermined Note: The effectiveness of the HIA cannot be determined by review of the HIA Report; this must be determined based on an internet/lit search.
Follow-up Measures	Monitoring, health impact management, or other follow-up measures called for in the HIA / N/A
Minimum Elements of HIA[5] Met? If no, what's missing	Yes / No - identify what's missing
GIS Used?	(If yes) Describe use – Illustrative, GIS analysis, etc. / No
Environmental/Ecosystem Impacts Considered?	(If yes) Identify impacts / No

---

[2] Harris, Harris-Roxas, Harris, and Kemp (2007)
[3] Human Impact Partners (2011)
[4] Wismar, Blau, Ernst, and Figueras (2007)

Table 1. Continued

Database Field	Description/Examples
Potential Improvements	Identify what could have potentially improved the HIA and/or its effectiveness. (Perhaps consult the HIA Practice Standards[5])
	Question to Consider: How are the HIAs different and what could have been done to close the gap?
	For example, quantification of impacts (including costing); consideration of environmental/ecosystem impacts; additional information; use of GIS/spatial analysis; broader utilization of existing tools/models/resources (C-FERST/T-FERST, BenMAP, National Atlas of Ecosystem Services, EJ View, MyEnvironment, UCLA Health Impact Decision Support Tool, etc.); consistency in conducting and reporting HIAs (e.g., sector terminology, enhanced guidance/methodology, transparent/publicly-accessible documentation); clear reporting of recommendations and mitigations; identification of evaluation and follow-up measures; etc.
Best Practices	Identify portions of the HIA process, report, etc. that stand out and describe these best practices.
	For example, tabular summary of potential impacts, including direction, extent, and populations most affected; defensibility of process; transparency of process documentation; etc.
	*Potentially identify a set of HIAs within each sector representing the best of the best.

---

[5] North American HIA Practice Standards Working Group (2010)

A Microsoft Access database was created using these finalized data entry fields to document the HIA Review. Included in this HIA Review Database was a table, a data entry form used to record the review of each HIA (Figure 4), and a report template that allowed database entries to be printed for review.

Final Literature Search

The preliminary literature search was updated in late spring 2012 to identify completed HIAs in the four sectors. A total of 91 completed HIAs were identified in this final literature search, however only 88 HIAs were available for review.

Full-scale Review

A team of five reviewers was enlisted to complete the full-scale review. HIA Review guidelines (*Appendix A*) were developed to provide the reviewers background on the project and more detailed guidance on review documentation and data entry. Prior to the start of the full-scale review, a training session was held to review the guidelines and prepare reviewers for the task.

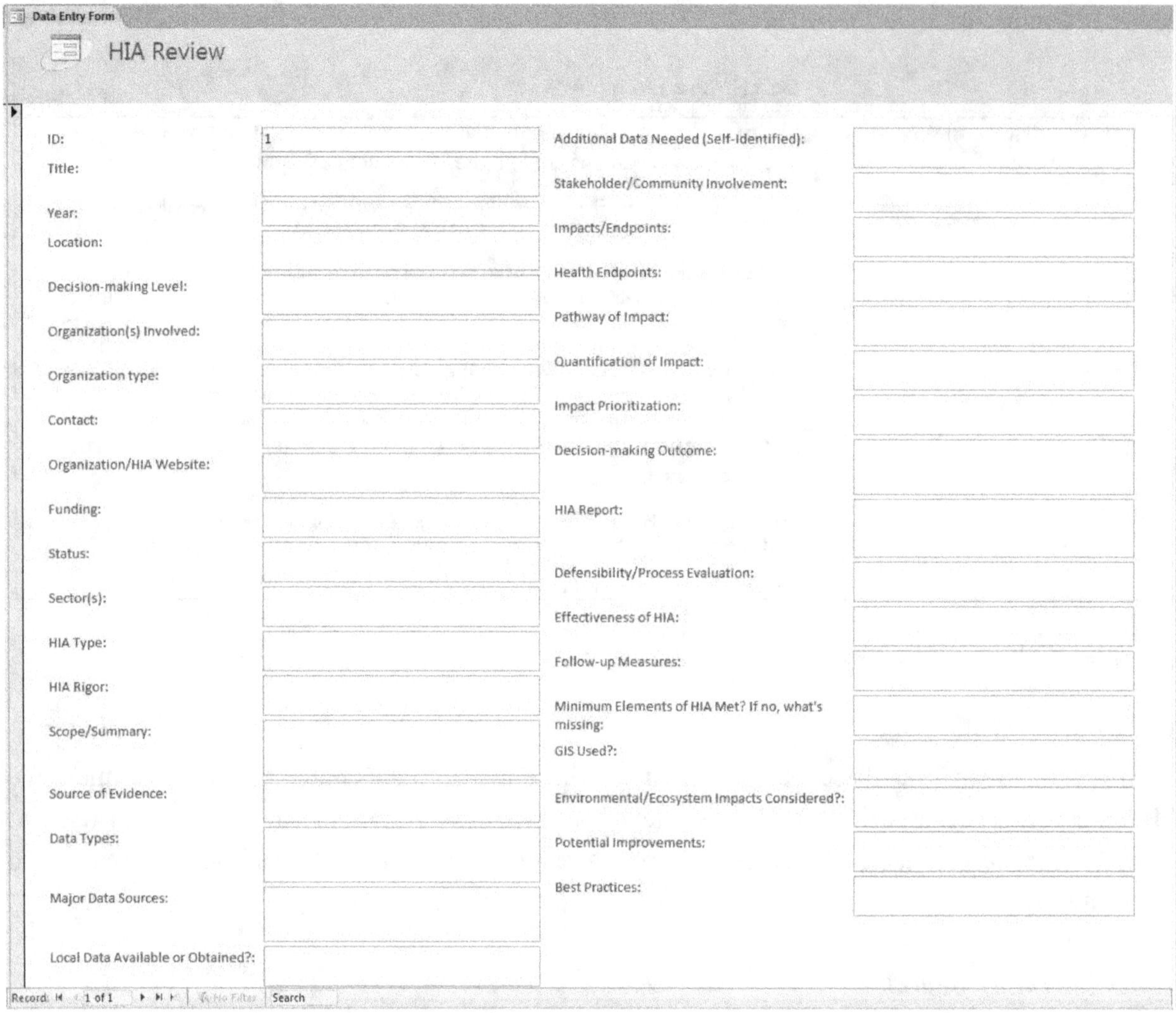

Figure 4. HIA Review data entry form.

Each of the five reviewers was assigned a set of HIAs to review and provided a copy of the Microsoft Access database for recording the results of the reviews. Data entry was based on each reviewer's independent review of the HIA and the guidance provided in the *Health Impact Assessment Review Guidelines* (*Appendix A*). Reviewers entered data directly into the data entry form of their database to populate the HIA Review Table; each HIA record was given a unique ID number.

Of the 88 HIAs obtained for review, 7 were removed from the review; 1 was found to be a duplicate of another HIA, 3 were found to not be true HIAs (i.e., one was a report on pedestrian collision modeling, one was a report on existing conditions, and one was a coordinated public transit-human services transportation plan), and 3 were Environmental Impact Assessments (EIAs) that did not report health impact assessment in a manner that allowed for analysis within the HIA Review framework (i.e., the HIA was integrated into the environmental impact statement [EIS] without a standalone HIA report). For further discussion on HIA and environmental impact assessment, including the integration of HIAs into EISs, see *HIA and Environmental Impacts*. At final count, 81 HIAs were reviewed from the Transportation, Housing/Buildings/ Infrastructure, Land Use, and Waste Management/Site Revitalization sectors – 4 in the pilot review and 77 in the full-scale review. See *Appendix B* for a list of the 81 HIAs reviewed and their respective sectors.

HIAs

Identified for review: 91

Unavailable for download: 3

Determined to be a duplicate: 1

Determined not to be an HIA: 3

Without standalone HIA report: 3

Final Number of HIAs Reviewed: 81

Quality Assurance

Quality Assurance (QA) measures were taken for both locating applicable HIAs for review and reviewing those HIAs. Following the preliminary literature search, an individual not previously involved in the project was asked to conduct a search for HIAs in the four designated sectors. The goal of this additional search was to determine whether any previously undiscovered, relevant HIAs existed. This review did not uncover any additional HIAs for review.

During the full-scale review, 10% of the HIAs underwent QA review to ensure consistency in the information being recorded. This QA review entailed select HIAs undergoing a second review by a person not involved in the initial review of the HIA and any discrepancies between the two reviews being examined and discussed. The QA review was conducted at the beginning of the full-scale review to set a standard for the remaining reviews. Eight HIAs across the four chosen sectors were designated for QA review and assigned a second reviewer.

Data entry from the initial and QA reviews of each of the eight selected HIAs was reviewed to ensure compliance with the database field descriptions, specific terminology, and/or formatting requirements established in the Review Guidelines, as well as general agreement in overall evaluation of the HIA. Some level of differences was expected in the evaluations due to subjectivity and level of detail used in recording the review. Data entry discrepancies between the initial review and the QA review of each HIA were identified and the discrepancies collectively reviewed to identify overall trends. Considerable discrepancies in data entry were found during the QA review, which could have been due, in part, to the unfamiliarity of the reviewers with HIAs in general, inexperience in applying the principles of the review process, or the implementation of the QA review early in the overall review process, as the HIAs in the QA review were the first HIAs examined in the full-scale review.

As a corrective action, a meeting was held with reviewers to discuss areas for improvement; challenges, questions, and lessons learned from the QA review; and the path forward. Areas for improvement included: use and consistency of specified terminology in designated fields and consistency in data entry, level of detail presented, and subjective evaluations. These areas for improvement are described in greater detail in *Appendix C*, along with feedback to reviewer challenges, questions, and lessons learned.

Reviewers were provided meeting notes documenting the proceedings of the corrective action meeting (*Appendix C*) and the initial and QA data entries for the HIAs they reviewed in the QA review. Reviewers were encouraged to go back and revise data entry for the HIAs in the QA review, as warranted. Midway through the full-scale review, one additional HIA was selected for QA review and reviewed by all five reviewers. The results of this second QA review showed marked improvement in data entry consistency.

It should be noted, however, that differences were still present in data entry, as was expected given that multiple people performed the reviews. To minimize these differences, a final QA check was conducted of all 81 data entry forms (see *Review Documentation*).

Review Documentation

At the completion of the full-scale review, the individual reviewer databases were compiled to form a single master database, and any duplicate data entry from the QA review was consolidated to provide one entry per HIA. Minor changes were made to the review framework during finalization of the master HIA Review Database to aid in analysis and more accurately reflect actual conditions. These changes included identifying a single (primary) sector for any HIAs originally identified by more than one sector; changing the "Quantification of Impact" field to "Characterization of Impact," since impacts in HIAs can be judged both qualitatively or quantitatively; and changing the "Impact Prioritization" field to "Prioritization Methods," since prioritization of impacts and recommendations were both recorded. A final QA check was conducted of each of the 81 data entry forms and edits made as needed to ensure compliance with the Review Guidelines and consistency in content, format, and level of detail throughout the database. This final HIA Review Database was subsequently analyzed to produce a synthesis of the HIA Review.

The final HIA Review Database will be available at: http://www.epa.gov/research/healthscience/health-review-hia.htm.

Health Impact Assessment Review Synthesis

The master HIA Review Database was analyzed to identify the general characteristics of the HIAs reviewed, the implementation and outcomes of the HIA process, and snapshots of HIAs in each sector.

General HIA Characteristics

This section of the synthesis examines the general characteristics of the reviewed HIAs, including status, title, year, location, and sector. The HIAs included in the HIA Review were those that fell within one of the four identified sectors, were complete, and had an HIA Report available for review (*See Appendix B* for a list of the HIAs reviewed).

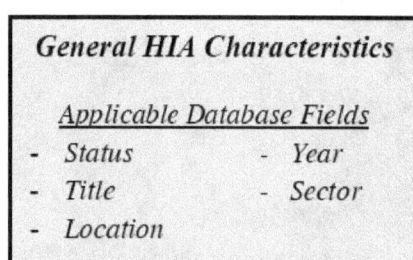

The reviewed HIAs were completed between 2005 and spring 2012, when the HIA Review began (Figure 5), at locations throughout the nation (Figure 6). Consistent with the overall trend noted by Health Impact Project (2013), the number of HIAs completed in these sectors is on the rise (Figure 5). Of the 81 reviewed HIAs, 48.2% were in the Land Use sector, 25.9% were in the Transportation sector, 21.0% were in the Housing/Buildings/Infrastructure sector, and 4.9% were in the Waste Management/Site Revitalization sector (Figure 6). The state of California has clearly been a leader in the HIA efforts in these four sectors, with over one-third of the reviewed HIAs being conducted there.

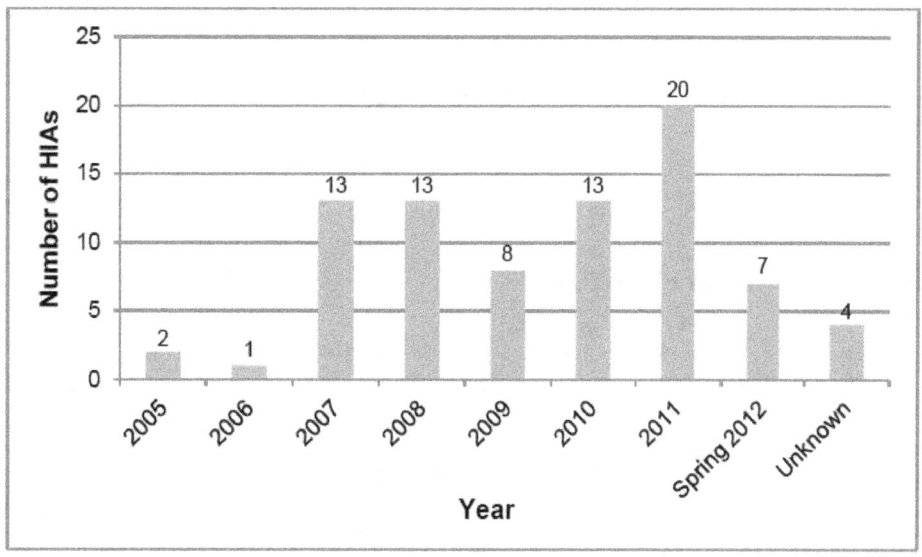

Figure 5. Year of report publication for reviewed HIAs.

13

HIAs by sector

- Transportation
- Housing/Buildings/Infrastructure
- Land Use
- Waste Management/Site Revitalization

# of HIAs completed in select sectors*

0   1   2   3   4   5   8   11   24
* 1 federal HIA

Figure 6. Number of reviewed HIAs completed by state and sector.

Implementation and Outcomes of the HIA Process

This section of the synthesis examines the HIAs in light of the six steps of the HIA process. The typical tasks involved in each HIA step are outlined in text boxes throughout the section using guidance from the North American HIA Practice Standards Working Group (2010) and the National Research Council (2011). Database fields related to each step were analyzed to provide a picture of the implementation and outcomes of those steps in the 81 HIAs reviewed. Note that some database fields are applicable to more than one step.

14

Screening

The screening step of the HIA process was not well documented in the HIAs that were reviewed. In fact, less than half of the HIAs (n=39) described undertaking the screening process at all.

Need for HIA to Inform Decision-making

When embarking on the HIA process, one of the first matters is to identify decisions under consideration by decision-makers and determine whether there is a need for HIA in those decisions. HIAs should be initiated when there is the potential for the HIA to add value to the decision-making process (e.g.,

> ### *Screening*
>
> Determine whether an HIA is needed and the value added. Tasks include:
>
> - Defining the decision and its alternatives
> - Evaluating the value of performing an HIA
> - Assessing the feasibility of conducting the HIA given the timeframe and available resources
> - Determining the willingness of partners and stakeholders to participate in the HIA
>
> *Applicable Database Fields*
> - *Decision-making Level*   - *HIA Type*
> - *Scope/Summary*   - *Funding*

where health is not already being considered, where disproportionate health consequences are likely, etc.) and should be initiated with enough time for the completed HIA to inform the decision. Of the 81 HIAs reviewed, 3 were not initiated to inform a specific decision under consideration by decision-makers and 1 was initiated in advance of a decision, but was not completed in time to provide input into the decision-making process.

Decisions Assessed in HIAs

The majority (56.8%) of the reviewed HIAs were used as part of local, community-level decisions, but many others were conducted to inform county, state, and federal decision-making processes (Figure 7). Table 2 identifies the general types of decisions that the HIAs informed at each of these decision-making levels. A more in-depth discussion of these decisions and their outcomes can be found in the *Sector Snapshots*.

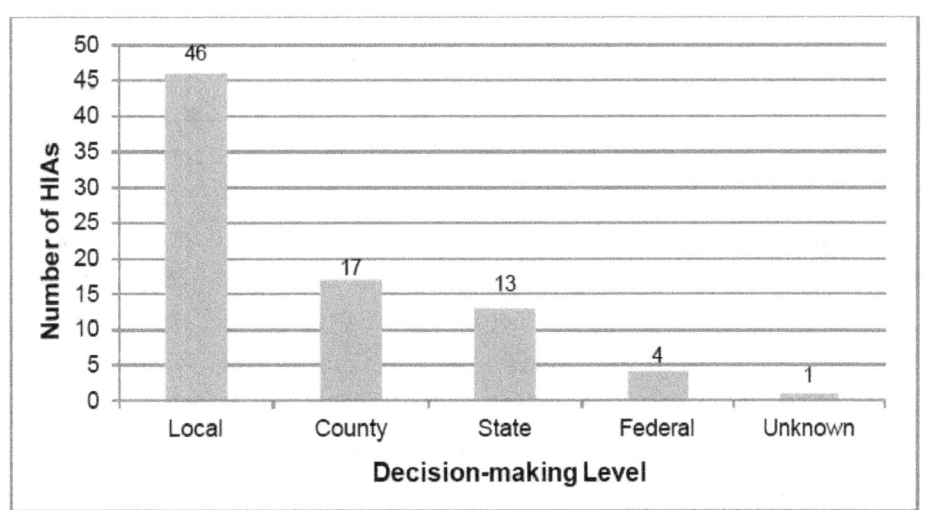

Figure 7. Number of review HIAs conducted at each decision-making level.

15

Table 2. Types of Decisions Informed by HIAs at Different Decision-making Levels

Local	County
- Transportation plans and policies - Redevelopment/restoration of historic districts - Alcohol policies - Land use policies and projects - Zoning controls and zoning code rewrites - Redevelopment/master plans - Mass transit/transit-oriented design - Pedestrian bridge projects - Siting of recreational centers and schools - Comprehensive plans - Growth policies - Road construction, redesigns, and infrastructure improvements - Rezoning plans and land repurposing - Remodels or expansion of community institutions (e.g., airports, hospitals, farmers markets) - Neighborhood/sub-area planning studies - Land use projects - City planning practices - Building demolition - Road pricing scenarios - Affordable housing siting - Port growth - Redevelopment of distressed public housing	- Siting of special uses (e.g., dirty materials recovery facility, biosolids storage facility) - County bicycle and pedestrian master plans - Placement and maintenance of community gardens - Growth alternatives - County plans and policies (e.g., agriculture, open-air burning) - Accessory dwelling unit (ADU) policies - Land zoning variances - Sub-area plans for revitalizing highways and surrounding neighborhoods - Bridge replacement projects - Proposed industry (e.g., coal-fired electric plants) - Natural gas development and production - Comprehensive/general plans and plan updates **State** - Mass transit and highway and bridge design - Housing and energy assistance programs - Comprehensive planning and growth policies - Energy programs and natural resource management, including fossil fuel exploration and development, and renewable energy and water management policies **Federal** - Oil and gas leases/developments (NEPA/EIS) - Federal housing policies

Types of HIAs Conducted

The HIAs were categorized into one of four types of assessment – decision-support, advocacy, mandated, or community-led – based on the details given in the HIA report and the descriptions provided for each of these typologies by Harris-Roxas and Harris (2011). As shown in Figure 8, decision-support HIAs are those conducted by or with agreement of the project proponents and/or decision-makers in order to improve the decision-making process; advocacy HIAs are those conducted by organizations that are neither part of the project nor the deciding body in order to bring under-recognized health concerns to light; mandated HIAs are those conducted to meet a statutory or regulatory requirement; and community-led HIAs are those conducted by the potentially-affected populations (Harris-Roxas and Harris 2011). With exception of five of the reviewed HIAs, all were either advocacy or decision-support HIAs (Figure 9); the remaining HIAs were mandated by legislative directive or NEPA (n=3) or were community-led (n=2).

Forms of health impact assessment			
	Mandated	Decision-support	Advocacy	Community-led
Description	Occurs in the context of an environmental impact assessment (EIA), integrated impact assessment (IIA) or environmental, social and health impact assessment (ESHIA) and is done to meet a regulatory or statutory requirement	Conducted voluntarily by, or with the agreement of, organisations responsible for a proposal, with the goal of improving decision-making and implementation	Conducted by organisations or groups who are neither proponents or decision-makers, with goal of influencing decision-making and implementation	Conducted by potentially affected communities on issues or proposals that are of concern
Purpose	• Meeting a regulatory or statutory requirement • Minimising negative health impacts	• Improving decision-making and implementation • Minimising negative health impacts • Maximising positive health impacts	• Ensuring under-recognised health concerns are addressed in design, decision-making and implementation • Minimising negative health impacts • Maximising positive health impacts	• Ensuring the community's health-related concerns are identified and addressed • Enabling greater participation of communities in decisions that affect them • Minimising negative health impacts • Maximising positive health impacts
Origins	Environmental health	Environmental health, social view of health, health equity	Social view of health, health equity	Social view of health, health equity
Role of values and judgements	Almost no role for values in assessment, judgements often not acknowledged	Implied role for values and judgements	More explicit role for values and judgements	Driven by community values and judgements
Conducted by	Consultants	Government agencies, consultants	Non-governmental organisations (NGOs), universities, other agencies	Communities, often aided by HIA practitioners in NGOs, universities or other agencies
Resourced by	Proponents	Government agencies	Varied	Communities themselves
Overseen by	Proponents	Government agencies	Varied	Communities themselves
Role of stakeholders	Providing technical information	Informing the assessment	Guiding the assessment	Controlling and conducting the assessment
Type of learning	Technical	Technical/conceptual	Conceptual/social	Social

Figure 8. HIA typology descriptions (Source: Harris-Roxas and Harris 2011).

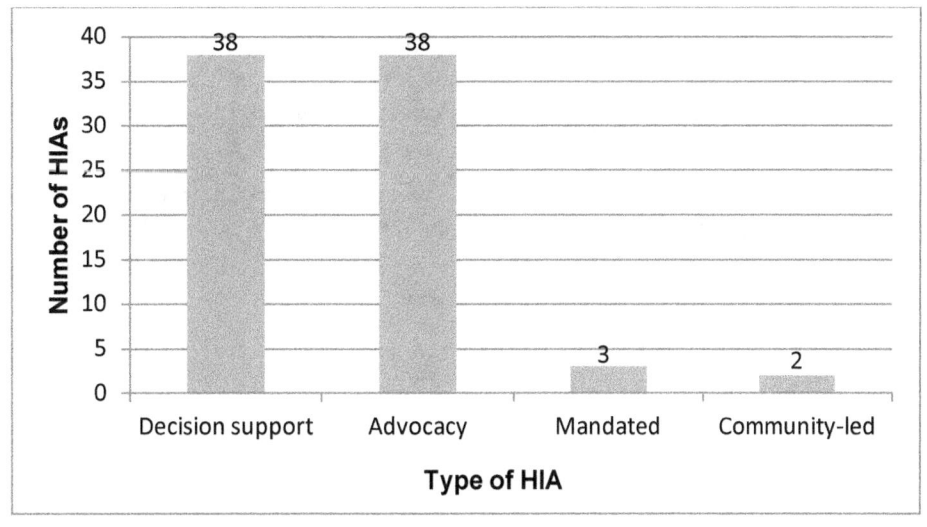

Figure 9. Types of HIA conducted.

HIA Funding

Sources of funding were unable to be determined for approximately 30% (n=24) of the HIAs reviewed, which could indicate that either: a) funding sources were not adequately documented, b) the HIAs were conducted within the scope of normal work activities (i.e., without any external funding), or c) the HIAs were performed by volunteers (e.g., a working group). Other HIAs were conducted with funding from one or more of these entities: the Centers for Disease Control and Prevention (CDC), the Robert Wood Johnson Foundation (RWJF), the Association of State and Territorial Health Officials (ASTHO), The California Endowment, the Health Impact Project, Human Impact Partners, the National Association of County and City Health Officials (NACCHO), Blue Cross/Blue Shield, the National Network of Public Health Institutes (NNPHI), the Northwest Health Foundation (NWHF), or others (Figure 10).

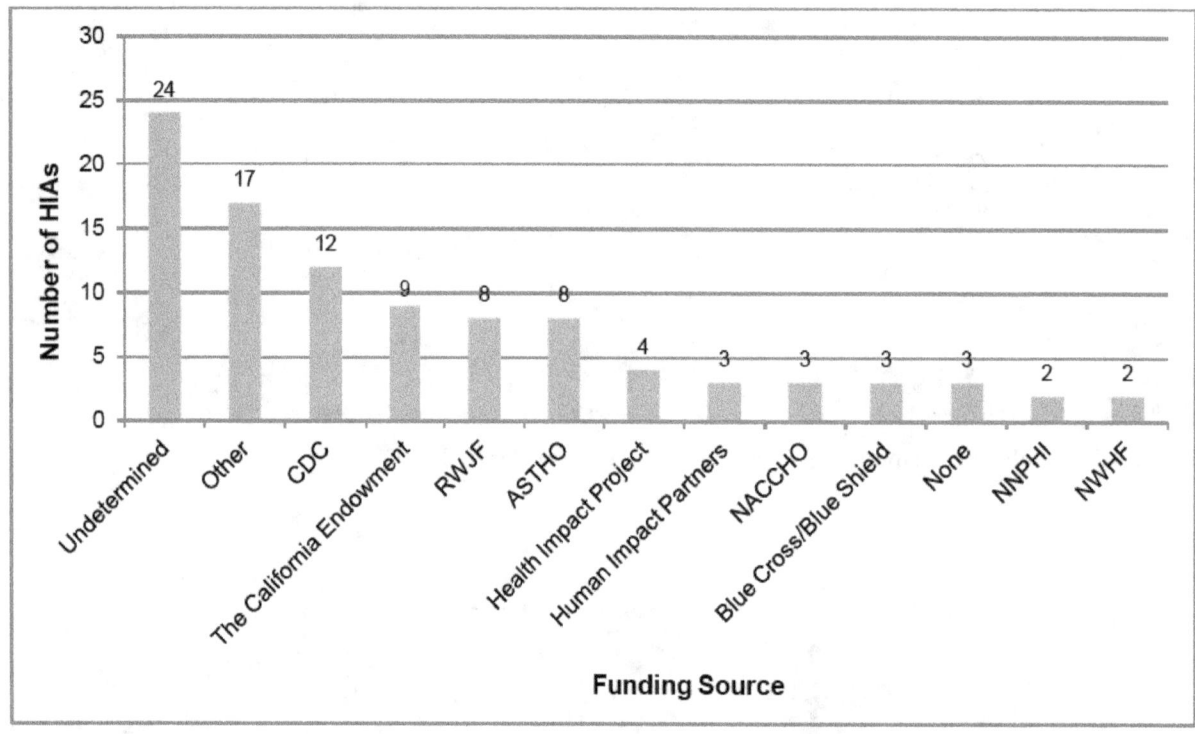

Figure 10. Funding sources of reviewed HIAs (CDC- Centers for Disease Control and Prevention; RWJF- Robert Wood Johnson Foundation; ASTHO- Association of State and Tribal Health Officials; NACCHO- National Association of County and City Health Officials; NNPHI- National Network of Public Health Institutes; NWHF- Northwest Health Foundation).

Scoping

Composition of HIA Teams

A number of different types of organizations were involved in conducting the reviewed HIAs (Figure 11), such as educational institutions, government agencies, non-profit groups, and others (e.g., consultants, research organizations, for-profit companies, health care companies, partnerships, working groups, etc.). Over half of the HIAs (n=44) were conducted by two or more of these entities. Of those 44 HIAs, almost 40% (n=16) were conducted by two government agencies or a combination of government agencies and non-profit organizations.

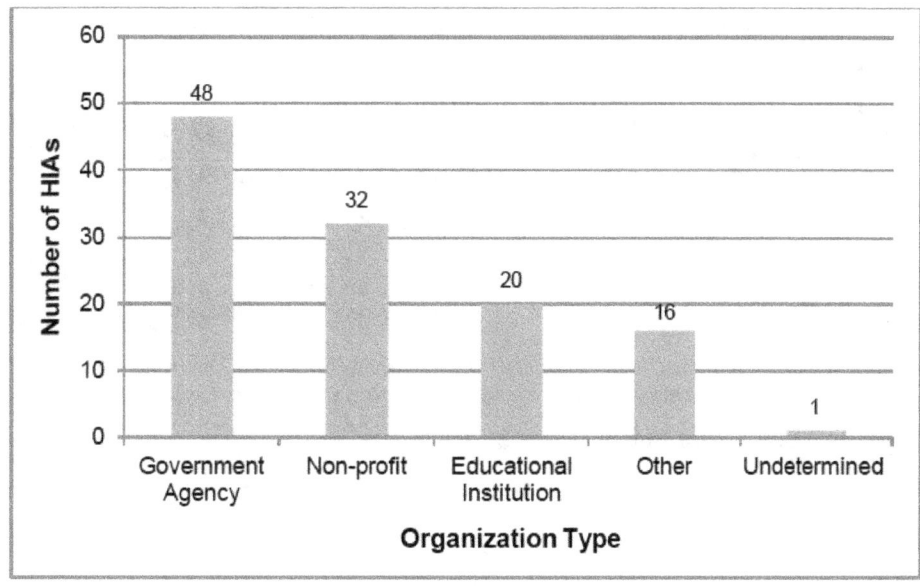

Figure 11. Types of organizations involved in conducting the reviewed HIAs.

Of the 37 HIAs conducted by a single organization, 20 were conducted by a government agency. State or local government health agencies were the most common organizations involved in conducting the HIAs, having their hand in 37 of the 81 HIAs. Human Impact Partners, a national non-profit involved in promoting health and equity in decision-making, conducted or was a partner in 13 of the HIAs reviewed.

HIA Level of Rigor

Part of the scoping process involves determining the rigor or level of HIA that will be conducted, including the number of impacts that will be assessed, the depth of assessment (e.g., extent of data collection, stakeholder involvement, sources of evidence, etc.), and the length of time that is available to complete the HIA. The rigor of the reviewed HIAs was judged using the definitions of four levels of HIA provided by Harris et al. (2007). These levels, listed from least to most rigorous (and least to most resource-intensive), are: desk-based, rapid, intermediate, and comprehensive (Figure 12). Figure 13 shows a breakdown of the number of HIAs performed at each level of HIA. It should be noted that reviewers also recorded in the database whether the rigor designation they assigned the HIA differed in any way from how the authors classified the assessment. There were ten cases of divergent classification, seven of which involved a rapid HIA being classified by reviewers as an intermediate HIA.

DESK BASED	RAPID	INTERMEDIATE	COMPREHENSIVE
No more than three impacts, assessed in less detail	No more than three impacts, assessed in more detail	Three to ten impacts, assessed in detail	All potential impacts, assessed in detail
Provides a broad overview of potential health impacts	Provides a more detailed overview of potential health impacts	Provides a more thorough assessment of potential health impacts, and more detail on specific predicted impacts	Provides a comprehensive assessment of potential health impacts
Is an 'off the shelf' exercise based on collecting and analysing existing accessible data.	Involves collecting and analysing existing data with limited input from experts and key stakeholders	Involves collecting and analysing existing data as well as gathering new qualitative data from stakeholders and key informants.	Involves collecting and analysing data from multiple sources (qualitative and quantitative)
2-6 weeks for one person full time[1].	6 to 12 weeks for one person full time.	12 weeks to 6 months for one person full time.	6 to 12 months for one person full time.

[1]The time involved will vary depending on the number of people actively involved in undertaking HIA tasks. For example, a comprehensive assessment may take a team of four people three months to complete.

Figure 12. Levels of HIA (Modified from Harris et al. 2007).

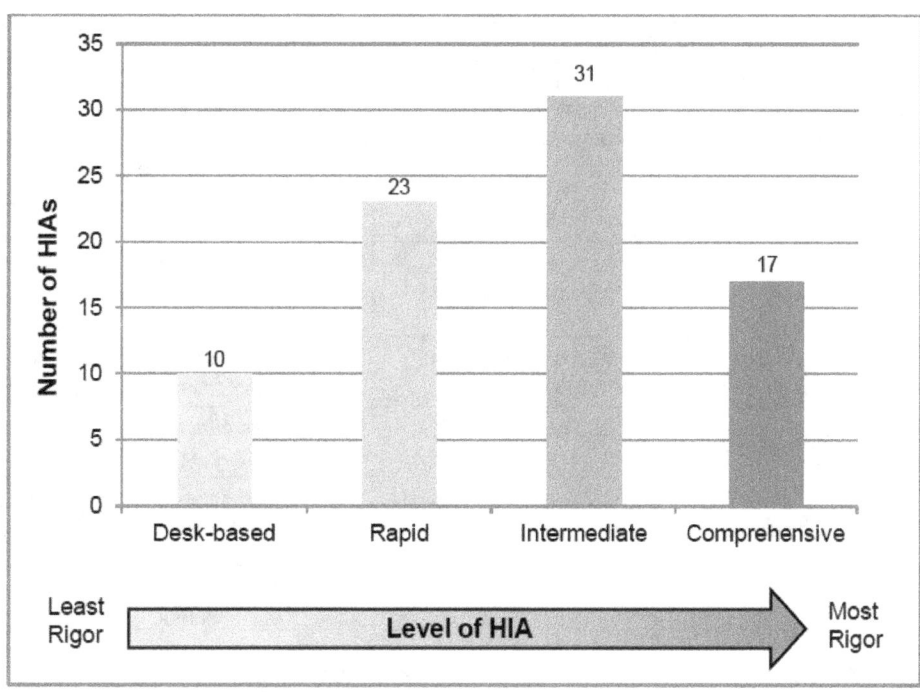

Figure 13. Rigor of reviewed HIAs.

Stakeholder Involvement

The rigor of the HIA determines the overall extent to which stakeholders will be involved throughout the HIA process. When establishing a plan for stakeholder involvement, it is important to consider not only who should be invited to participate in the process (e.g., residents, community-based organizations, decision-makers, government leaders and representatives, business and industry, advocacy organizations, academic institutions, policy and subject matter experts, etc.), but also the level of engagement and methods of stakeholder participation to be used (North American HIA Practice Standards Working Group 2010; National Research Council 2011). The level of stakeholder engagement can range from input to empowerment, as described in the Ladder of Citizen Participation (Arnstein 1969; Figure 14), and methods of stakeholder participation can include avenues such as steering or advisory groups, needs assessments, community listening sessions, public comment periods, interviews, surveys, project meetings, focus groups, expert consultations, and forums or workshops, among other things (Stakeholder Participation Working Group 2010).

Sixty-six (66) of the 81 HIAs had some type of stakeholder and/or community involvement component. This is not surprising given that stakeholder engagement and community empowerment are objectives of HIA (North American HIA Practice Standards Working Group 2010; National Research Council 2011; Human Impact Partners 2012).

Rung	Arnstein's (1969) Description	Applied to HIA Practice
Citizen Control & Delegated Power	_Vulnerable populations_* most impacted obtain majority decision-making power.	HIA stakeholders, including vulnerable populations, decide on the HIA scope and recommendations, have final approval of HIA report, and decide on the communications strategy.
Partnership	Vulnerable populations can negotiate and engage in trade-offs with power holders.	Stakeholders impact the direction of HIA (scope) and reporting, but decisions are made equally with project team.
Placation	Allows vulnerable populations to advise, but power holders have right to decide.	Stakeholders offer input that may shape the HIA, but the project team make all decisions.
Informing & Consultation	Citizens can offer input and be heard, with no assurance their views will be taken into account.	Stakeholders offer input but it does not necessarily shape the HIA.
Manipulation & Therapy	Power holders -educate or -cure citizens— participation is not encouraged.	Telling stakeholders what is happening without soliciting input. Saying stakeholder voices matter but not acting on input. Not giving out all relevant information or giving different information to different stakeholders.

* Authors took liberty to change Arnstein's use of the term "have not" and replace it with "vulnerable population."

Figure 14. Ladder of Citizen Participation in HIA (Source: Stakeholder Participation Working Group 2010).

In fact, HIA guidance calls not only for the participation of the community and stakeholders affected by the decision, but also the decision-makers themselves. Although the HIA Review recorded, to the extent possible, the stakeholders involved in each HIA and the method(s) of engagement and/or participation, it did not specifically examine the inclusion of decision-makers as stakeholders in the HIA process. Nevertheless, reviewers explicitly identified decision-makers as stakeholders in a handful of HIAs (n=3). Decision-makers may have been engaged as stakeholders in additional HIAs as well, but this was not evident from a cursory review of the HIA Review Database.

Among the HIAs with a stakeholder or community involvement component, the level and quality of stakeholder participation varied greatly. In many of these HIAs, stakeholder input was solicited via interviews, surveys, public meetings, community forums and workshops, and/or other special collection methods to inform the scoping step (e.g., identify issues of interest and areas of concern for the community and stakeholders, identify populations and vulnerable groups that might be affected by the decision, etc.) and gather local knowledge regarding community health and existing conditions to inform the assessment step of the process. Public or project meetings were not only used for soliciting input from stakeholders, but were also a common method used for communicating the results of the assessment and recommendation steps of the HIA process to stakeholders and in some instances, soliciting their feedback and comments. Often, stakeholders were only minimally engaged in the process (i.e., not involved in the actual HIA decision-making), but there were a number of HIAs (n=15) that engaged stakeholders in the decision-making process, usually via a role on an advisory or steering committee, although other methods were also used (e.g., stakeholder panels and councils, meetings, and forums). In a handful of HIAs (n=4; two community-led HIAs and two advocacy HIAs), stakeholders actually oversaw or guided the HIA process and were engaged as decision-makers in equal partnership with the HIA team or as the primary decision-makers in the process.

Of the 15 HIAs with no stakeholder and/or community involvement, ten were desk-based HIAs, which by definition do not include stakeholder or community involvement, and the remaining HIAs incorporated stakeholder and/or community input gathered outside of the actual HIA process (i.e., there was no stakeholder or community involvement in the HIA itself, because the HIA used previously-collected stakeholder data).

Types of Impacts Identified for Assessment

Another important part of the scoping process involves determining what impacts will be assessed in the HIA. This is often first accomplished by determining the impact the decision could have on known determinants of health, such as individual factors; individual behaviors; public services and infrastructure; living and working conditions; and social, economic, and political factors (Figure 15). These determinants of health are factors known to directly or indirectly impact an individual's health. Oftentimes, it is not feasible, or even possible, to examine all of the impacts of a decision (i.e., perform a comprehensive HIA). In these cases, a determination needs to be made as to which impacts to include in the HIA. While this decision is usually established early in the scoping process by the HIA team and/or stakeholders, the impacts chosen to be examined can be revised through stakeholder input later in the scoping step and even through research, stakeholder input, and analysis in the assessment step of the HIA process.

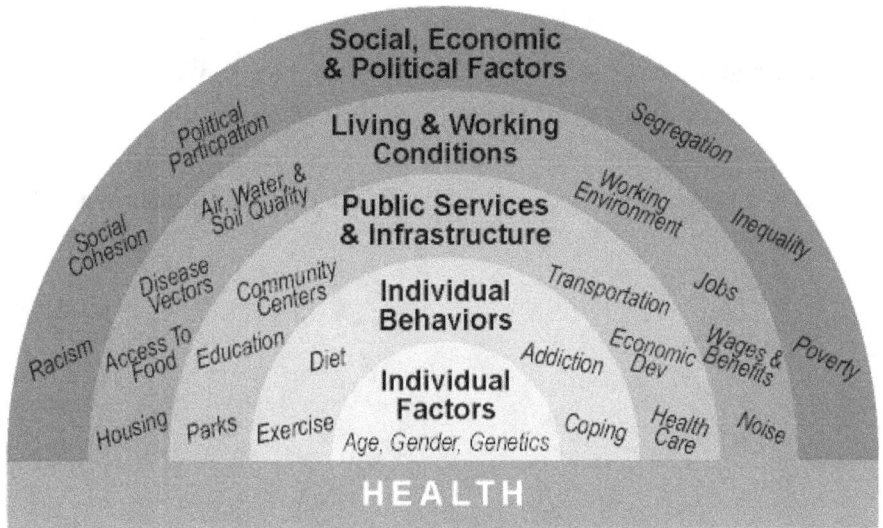

Figure 15. Determinants of health (Source: Human Impact Partners 2011).

23

Prioritization of impacts can be based on a number of factors, including stakeholder/community input; distribution/equity of impacts; literature/research; impact on health; direction of impact; duration/timing of impact; geographic extent of impact; relevance to project/decision interests; likelihood, magnitude, and permanence (i.e., severity) of impacts; measurability of impact; data availability/data gaps; quality of evidence; consultation with experts; population affected; or specific prioritization criteria or ranking systems implemented for the project. Figure 16 shows the methods identified most frequently in the reviewed HIAs (i.e., in ten or more of the HIAs) for prioritizing impacts.

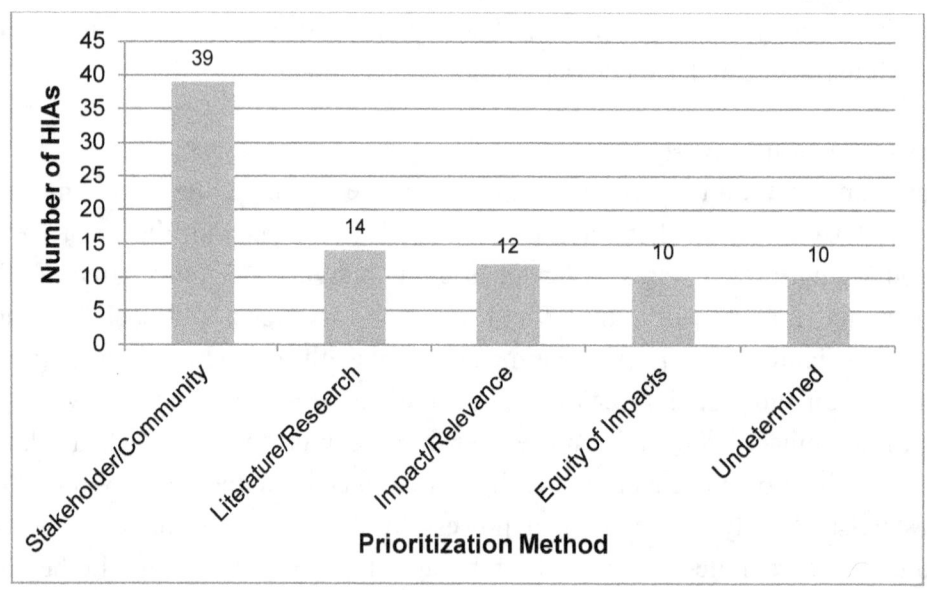

Figure 16. Methods used to prioritize impacts in reviewed HIAs.

The impacts (or endpoints) assessed in each HIA were classified by reviewers into one of the following categories: health (physical, mental, developmental), environmental/ecosystem (e.g., impacts on the natural environment; air, water, and soil quality; noise pollution), behavioral, economic, infrastructure (e.g., built environment), services, demographic, or other (Figure 17). These categories relate to the generally-accepted determinants of health shown in Figure 15.

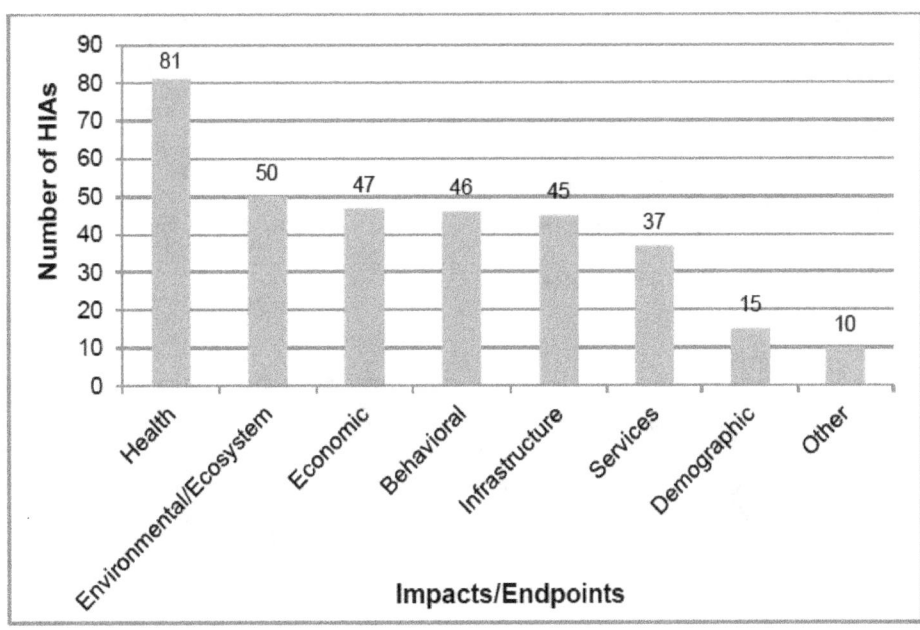

Figure 17. Impacts/endpoints assessed in reviewed HIAs.

As would be expected, all 81 HIAs examined health endpoints. The endpoints classified as other included educational attainment, cultural, social, and spiritual impacts. Environmental/ ecosystem endpoints were the second most common type of endpoint examined (n=50), primarily due to the frequency of air quality impact assessments. Air quality impacts were assessed in over half (n=42) of the HIAs reviewed (Figure 18) and commonly involved, in whole or in part, examination of traffic-related impacts on air quality. This result may be due, in part, to the availability of a number of models and forecasting tools from EPA and others for predicting traffic-related emission and pollutant dispersion rates.

The frequency in which other environmental/ecosystem impacts were examined in the HIAs is shown in Figure 18 and included water quality, green/open space, vegetation, wildlife, climate, soil, habitat (e.g., habitat quality, loss, and fragmentation), greenhouse gases, environmental stewardship, and ecosystem management impacts.

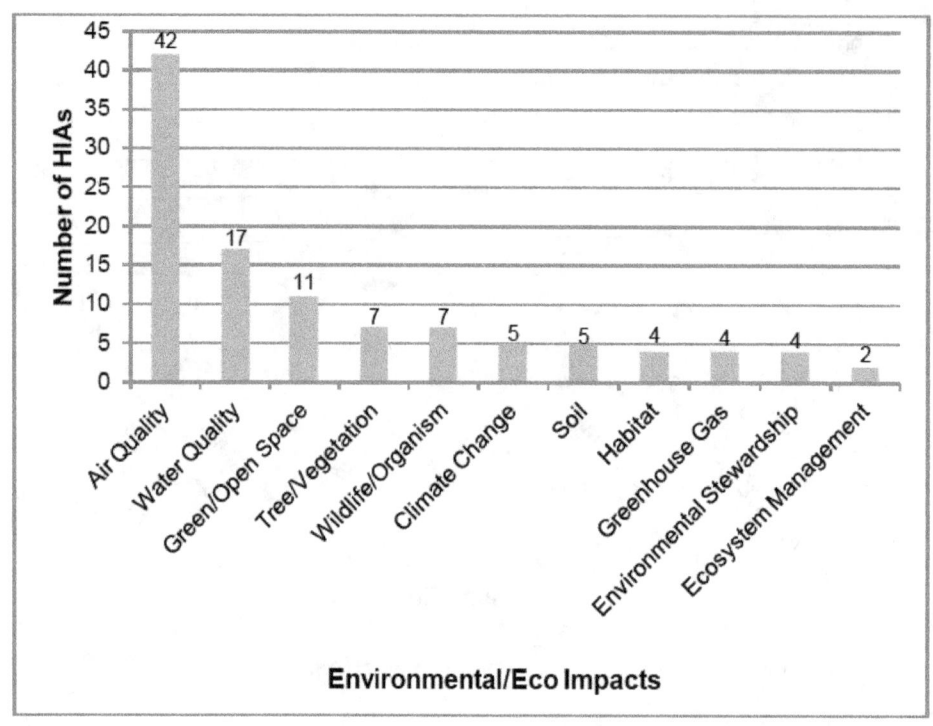

Figure 18. Environmental and ecosystem impacts assessed in reviewed HIAs.

Pathways of Impact

To determine the potential health endpoints or outcomes of the impacts assessed in HIAs, the pathways through which those impacts will occur must be identified. Pathway diagrams are a common tool used to illustrate the links between chosen health determinants and expected health outcomes. Figure 19 is a detailed depiction from the HOPE VI to HOPE SF San Francisco Public Housing Redevelopment HIA (UCBHIG 2009) showing the pathways between housing and health at the social, macroenvironmental, and microenvironmental scales. Logic frameworks can also be developed to show the various, interconnected pathways of impact for a decision, since the health impacts of a decision are not likely to occur through a single pathway. Figure 20, taken from the Health Effects of Road Pricing in San Francisco HIA (SFDPH 2011) shows the various pathways through which road pricing policies may affect health.

Housing

	Social → Macroenvironmental → Microenvironmental		

Housing factors

Deprivation	Neighborhood	Design	Maintenance
• Homelessness • Affordability • Neighborhood housing quality inequities	• Crime and violence • Traffic • Access to healthy amenities • Collective efficacy	• Heating Ventilation Air Conditioning (HVAC) • Water supply • Materials • Structural integrity • Lighting • Stairways • Common areas	

Determinants

• Overcrowding • Economic trade-offs	• Safety • Noise • Air pollution • Sidewalks, greenspace, recreation, schools, work, shopping, food environment	• Lead exposure • Mold exposure • Pest infestation • Exposure to allergens • Temperature extremes • Air pollution • Exposure to radon	• Exposure to asbestos • Exposures to toxic chemicals • Dirty carpets • Social interaction • Safety

Health outcomes

• Infectious disease • Poor nutrition	• Mortality • Physical activity • Cardiovascular health • Depression • Infectious disease • Poor nutrition	• Neurodevelopment • Mortality • Physical activity • Cardiovascular health • Depression • Infectious disease • Stress	• Asthma and respiratory health • Cancer • Heat/cold related illness • Injuries

Figure 19. Pathways between housing and health (Source: HOPE VI to HOPE SF San Francisco Public Housing Redevelopment HIA).

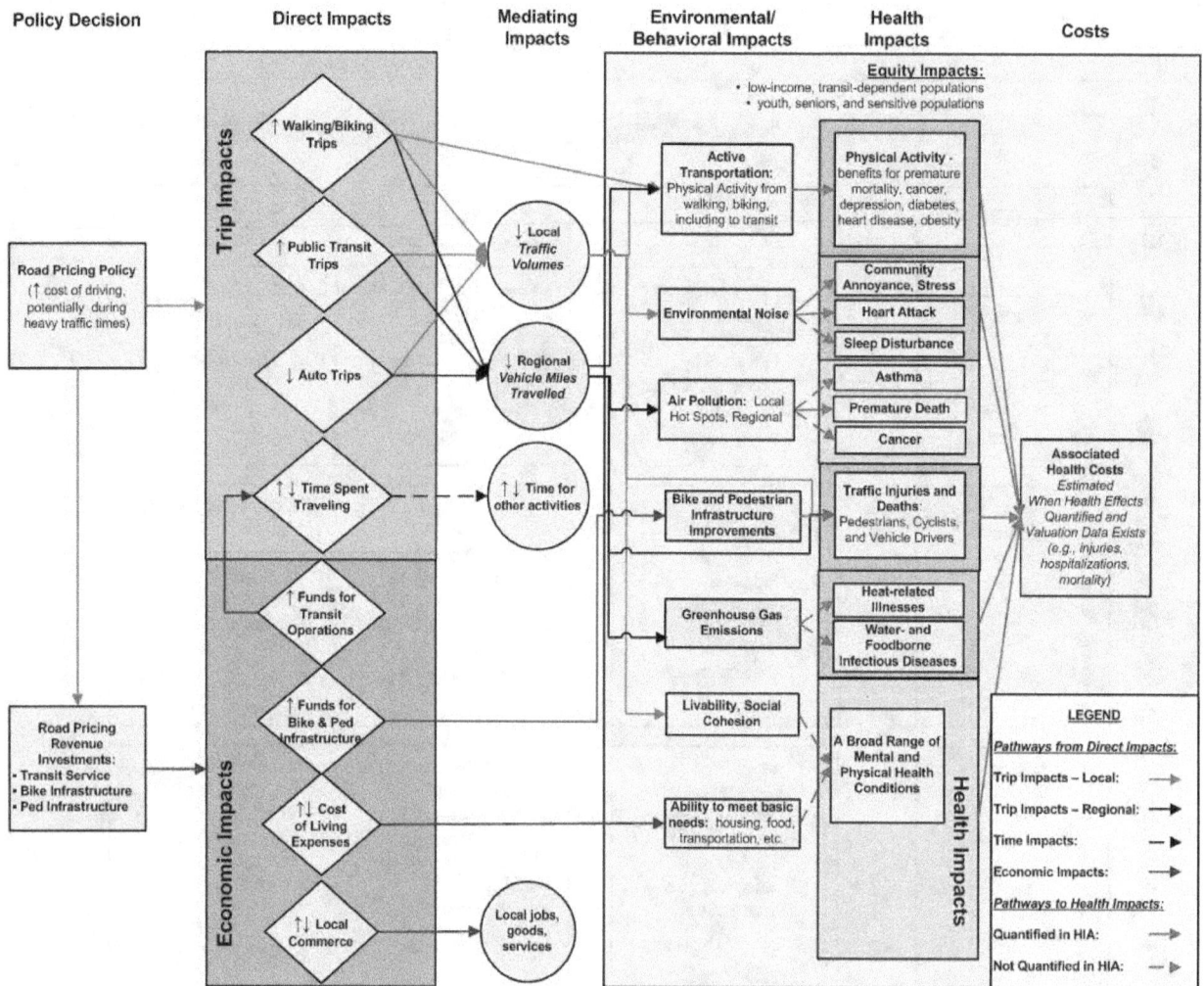

Figure 20. Logic framework showing pathways between road pricing and health (Source: Health Effects of Road Pricing In San Francisco, California: Findings from a Health Impact Assessment).

The pathways examined in the reviewed HIAs ranged from common pathways with well-studied human health effects, such as safety and security, mobility/access to services, physical activity, social capital, air quality, community and household economics, nutrition, exposure to hazards, land use, noise, housing, parks and recreation, education, healthcare access, and water quality, to less common pathways, such as culture, climate change, visual effects, habitat, and public participation (Figure 21).

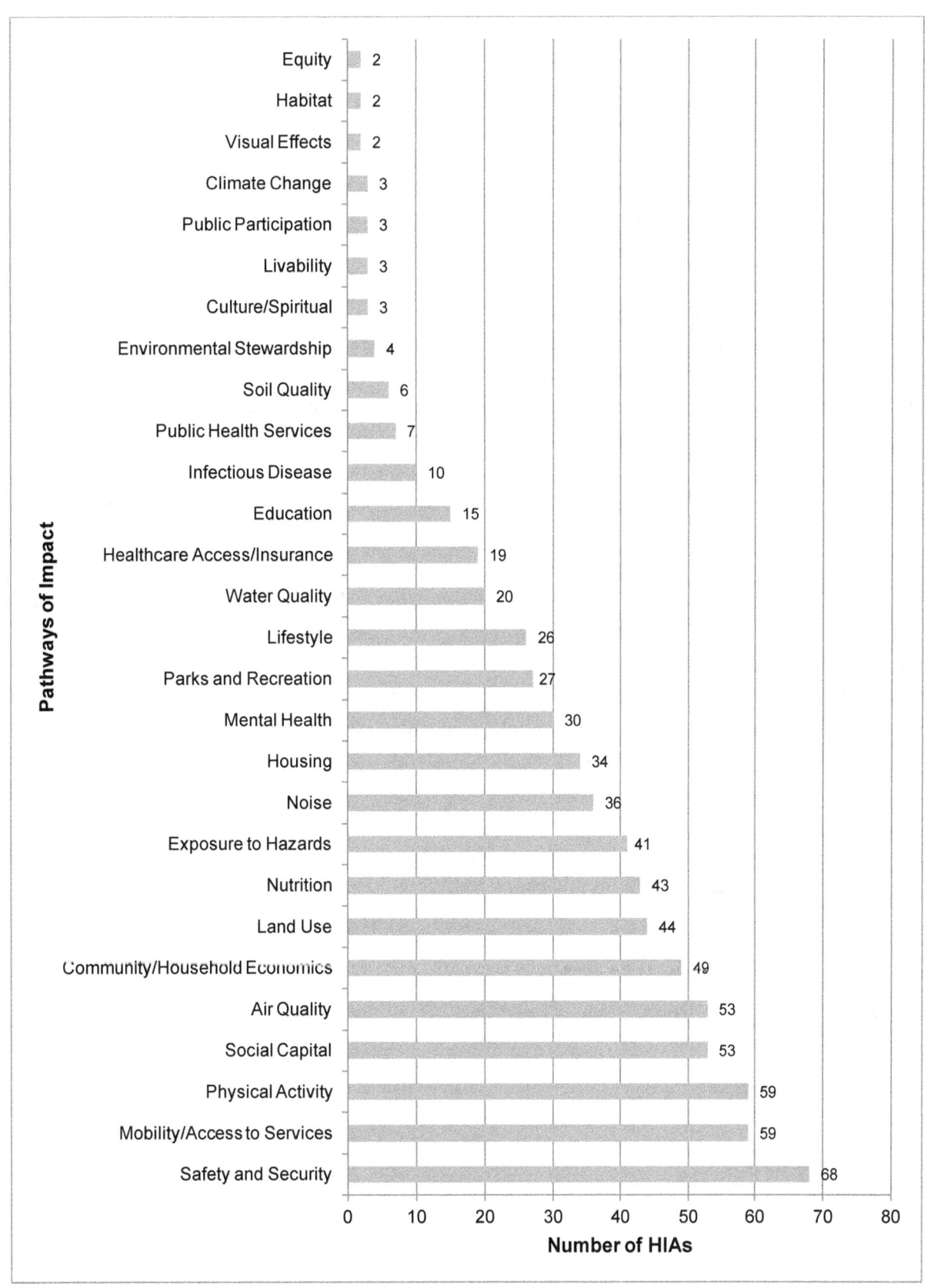

Figure 21. Pathways of impact assessed in two or more reviewed HIAs.

29

Like sector terminology, this is another area of the HIA practice that lacks an established set of terminology. Several common pathways examined in HIA have considerable overlap and/or subtle differences between them, leading to the terminology for these pathways being used interchangeably at times (e.g., transportation vs. mobility/access to services or land use vs. built environment or infrastructure). In this HIA Review, the mobility/access to services pathway was used to examine transportation-related impacts, and the land use pathway was used to examine impacts from both land use planning (e.g., zoning and siting of land uses) and physical land use (e.g., agriculture, industry, development, built environment/infrastructure, etc.).

HIA Scope and General Approach
In addition to identifying the organizations to be involved in the HIA, the rigor of the HIA, the level of stakeholder involvement, and potential impacts and pathways, the scope of the HIA (i.e., HIA research questions and goals) is also established in this step of the HIA process, as well as the general approach and depth of assessment that will be undertaken.

The scope of each of the HIAs reviewed is shown in its respective *Sector Snapshot*. The approach and depth of assessment to be undertaken takes into consideration the timeline for the HIA, resource availability, potential data sources and data gaps, and sources of evidence (i.e., methods). Sources of evidence can vary from literature and policy reviews to more resource-intensive community consultation (i.e., gathering information on community concerns) and special collection methods (e.g., expert consultation, forecasting, interviews, focus groups, modeling, risk assessment, and new data collection and analysis), as defined in Harris et al. (2007). The sources of evidence employed in the reviewed HIAs are identified in the Assessment section that follows (see Sources of Evidence and Data Types).

Assessment
The assessment step of the HIA process involves using data, tools, and methods to create a profile of existing health conditions and characterize the potential impacts of the decision.

Sources of Evidence and Data Types
Figure 22 shows the sources of evidence employed in the reviewed HIAs. The data types obtained through those sources included literature, existing and new data, websites, and models (Figure 23).

Assessment

Collect qualitative and quantitative information to create a profile of existing health conditions, and identify, evaluate, and prioritize the potential health impacts of the decision. Tasks include:

- Gathering existing data and collecting new data as needed; utilizing diverse sources
- Using data and existing tools and methods to profile existing conditions and evaluate potential health impacts of the decision
- Considering direction, magnitude, severity, likelihood, and distribution/equity of impacts via qualitative and quantitative analysis
- Describing data sources and methods used, including documentation of stakeholder engagement
- Acknowledging assumptions, strengths, and limitations of data and methods used

Applicable Database Fields

- Sources of Evidence	- Stakeholder/Community Involvement
- Data Types	- GIS Used?
- Major Data Sources	- Health Endpoints
- Local Data Available or Obtained?	- Characterization of Impact
- Additional Data Needed (Self-Identified)	- Defensibility/Process Evaluation

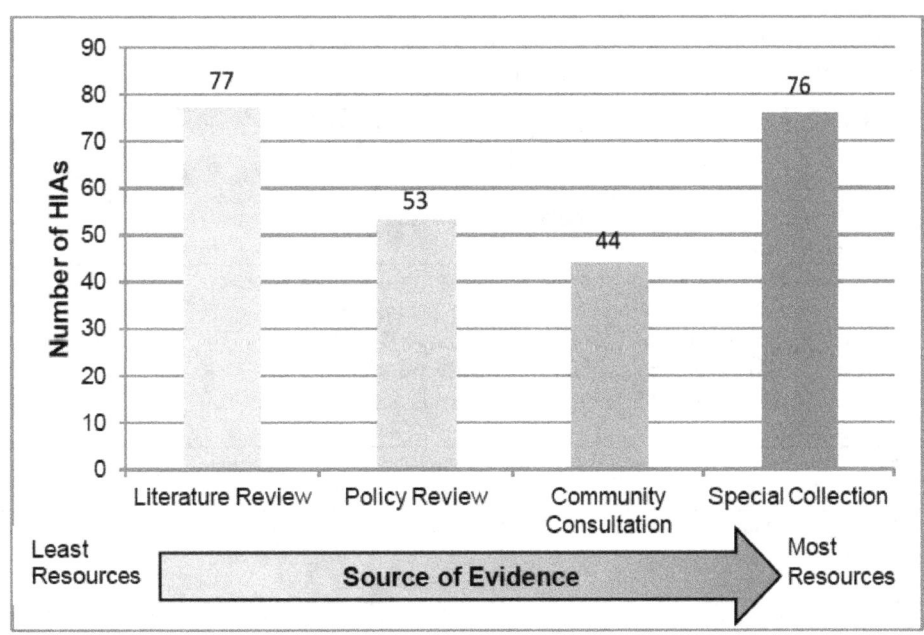

Figure 22. Sources of evidence used in reviewed HIAs (as defined in Harris et al. 2007).

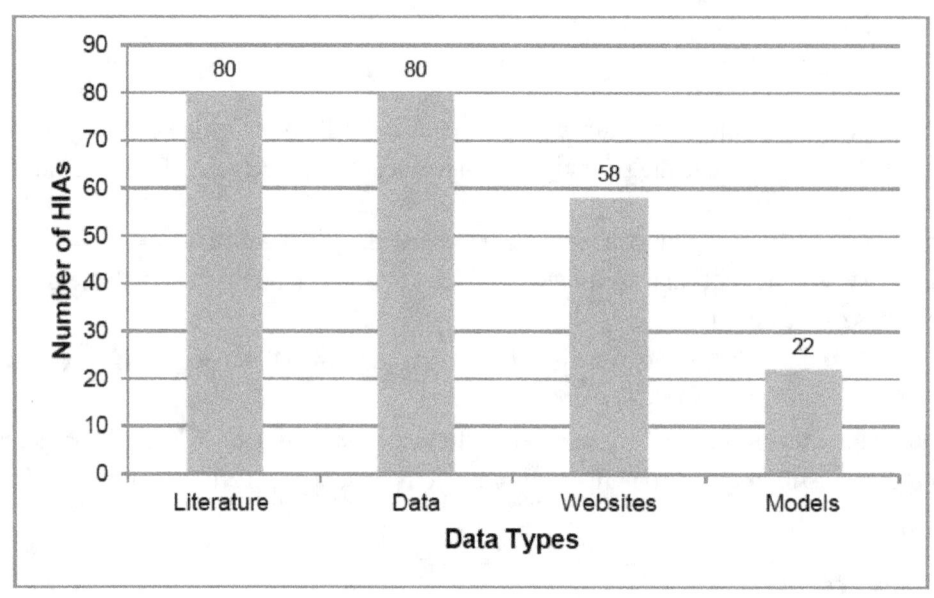

Figure 23. Data types used in reviewed HIAs.

Literature/Policy Review

As evidenced by Figures 22 and 23, gathering information from the existing literature is a significant component of HIAs. Table 3 shows the bibliographic resources identified in the reviewed HIAs for gathering evidence on pathways, health impacts, and endpoints.

Table 3. Bibliographic Resources Used in HIA Literature Reviews (resources shown in *italics* are publicly available, although access to actual publications varies by resource)

Bibiographic Resource	Description	Source
Cochrane Library	A collection of databases containing medicine and healthcare-related information.	http://www.thecochranelibrary.com/view/0/index.html
FirstSearch/ *WorldCat*	Available at a cost and offers web access to full-text articles, electronic books and journals, digitized special collections, etc. Databases are also visible at no cost through WorldCat.org, which allows users to search library collections around the world for an item of interest and then locate a nearby library that owns it.	http://www.oclc.org/firstsearch.en.html; http://www.worldcat.org/
Google Scholar	Provides a search of scholarly literature across many disciplines and sources, including theses, books, abstracts, and articles.	http://scholar.google.com/
Human Impact Partners (HIP) Evidence Base	A searchable database that includes research evidence and citations linking social determinants, the built environment, and health.	http://www.humanimpact.org/evidencebase
JSTOR	A digital library of more than 1,500 academic journals, books, and primary sources.	http://www.jstor.org/
LexisNexis Academic	Provides access to government and legal information, including government and political news, legal news, law reviews, and state and federal statutes and case law.	https://www.lexisnexis.com/hottopics/lnacademic/

Table 3. Continued

Bibiographic Resource	Description	Source
MEDLINE/ *PubMed*	Contains journal citations and abstracts for biomedical literature from around the world. PubMed provides free access to MEDLINE and links to full text articles when possible.	http://www.ncbi.nlm.nih.gov/pubmed
Ovid/Ovid MEDLINE	Ovid is a medical research platform that allows users to search content and productivity tools. Ovid MEDLINE is a comprehensive biomedical database that is updated daily and offers access to bibliographic citations and author abstracts from more than 5,500 biomedicine and life sciences journals.	http://www.ovid.com
ProQuest (formerly CSA Illumina)	Provides citations and abstracts for peer-reviewed journal articles, books, chapters and essays, dissertations, and more. Users can select a subject area or search across all subjects.	http://search.proquest.com/
PsycINFO	An abstracting and indexing database of peer-reviewed literature in the behavioral sciences and mental health.	http://www.apa.org/pubs/databases/psycinfo/index.aspx
Science Direct	A full-text scientific database offering access to journal articles and book chapters.	http://www.sciencedirect.com/
Scopus	World's largest abstract and citation database of peer-reviewed literature and includes tools that track, analyze, and visualize research.	http://www.scopus.com/
TRID/TRIS (Transportation Research Information Services)	An integrated database that combines the records from the Transportation Research Board's (TRB's) Transportation Research Information Services (TRIS) Database and the Organisation for Economic Cooperation and Development's (OECD's) International Transport Research Documentation (ITRD) Database. Provides access to transportation research worldwide.	http://trid.trb.org/
Web of Knowledge/ Web of Science	Can be used to access journal articles, patents, websites, conference proceedings, and Open Access materials in the sciences, social sciences, arts, and humanities. Web of Science can be found within Web of Knowledge and offers access to journal articles in the sciences, social sciences, and arts and humanities.	http://apps.webofknowledge.com/

Data Sources

Data, tools, and models used in the assessment phase of the reviewed HIAs were gathered from a variety of resources. The U.S. Census Bureau was the primary resource used in the HIAs for demographics and background data (e.g., social, economic, housing, and educational attainment data), while health data was most commonly gathered from state, county, or local health departments or one of the various health surveys conducted by the CDC. Figure 24 shows the most commonly used resources (i.e., the resources and/or organizations drawn upon in approximately one-quarter or more of the reviewed HIAs).

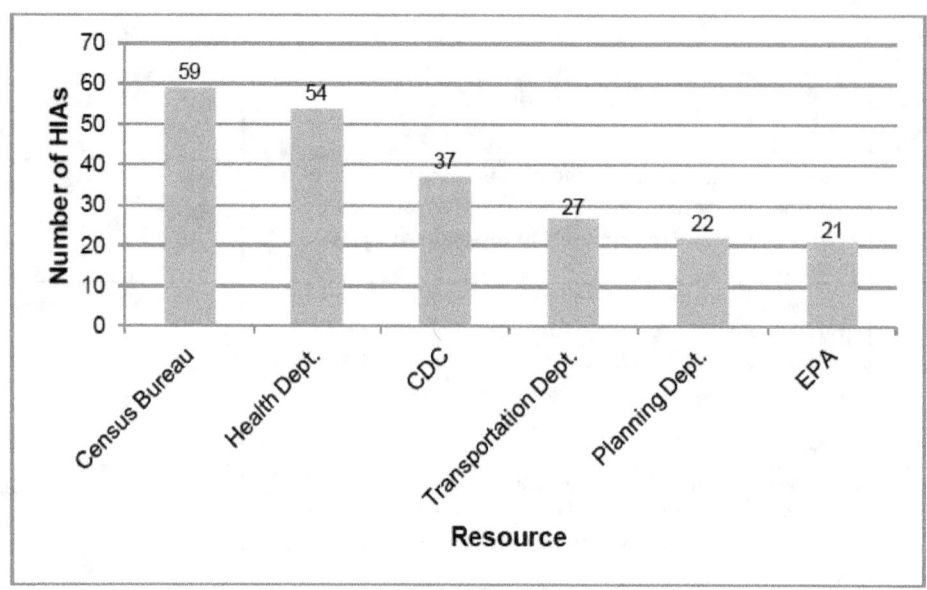

Figure 24. Most common resources drawn upon in reviewed HIAs for data, tools, and models (CDC- Centers for Disease Control and Prevention; EPA- Environmental Protection Agency)

The data utilized from these and other resources can be found in *Appendix D*. This list is not comprehensive, but does identify national data sources used in the HIAs and examples of data sources that may be available at the state, county, or local level. The data sources are organized into five categories: Demographics and Background Info, Health Data, Other Supporting Data, Benchmarks, and Indicators. Tools and models utilized in the HIAs can be found in *Appendix E*, including a description of the tool or model and its source.

Primary Data Collection

In addition to utilizing existing data, tools, and models, a variety of special collection methods were used in the HIAs to acquire new data, often at the "local" level or level of the project or decision. Table 4 shows a list of special collection methods used in the reviewed HIAs.

Table 4. Special Collection Methods Used in Reviewed HIAs

Special Collection includes:	
- accessing unpublished data	- health surveys
- advisory committee	- mediation group
- aerial photography	- modeling/forecasting
- air quality study	- PhotoVoice/community photography
- applicant information	- public/project meetings
- community forums/workshops	- residents panel
- demographics analysis	- risk assessment
- expert consultations	- stakeholder interviews
- field visits/site observations	- community/stakeholder surveys
- focus groups	- threshold scoring
- food audit/retail food availability survey	- traffic assessment/counts
- Geographic Information Systems (GIS) and photo mapping	- walkability audits
- Healthy Development Measurement Tool (HDMT)	- windshield surveys/tours

Stakeholder/Community Engagement

Of the utilized special collection methods, community forums/workshops, expert consultations, focus groups, health surveys, community photography, public/project meetings, residents' panels, stakeholder interviews, and community/stakeholder surveys were used most often as a means to involve and/or solicit information from stakeholders and the community.

Geospatial Analysis

GIS and other mapping techniques were also utilized quite frequently, not only for visualizing geographically-referenced data, but also in analysis to reveal geospatial relationships, patterns, and trends. Of the 54 HIAs that employed this special collection method, 42 utilized GIS in an illustrative capacity to display data and 29 utilized GIS in actual geospatial analysis.

Data Gaps

While all of the HIAs were able to obtain some type of local data for use in their analyses, there were instances where certain data were not available at the desired geographic scale or not available at all. In many instances, the authors of the HIA reports acknowledged these data gaps and identified the additional data needs. A summary of those data gaps is shown in Table 5, grouped by data category (i.e., baseline, employment sector/economic, environmental, health, causality, infrastructure/services, miscellaneous, program, local/small geographic scale, temporal, permit/application/plan, and tools) and a complete list of the HIAs' self-identified data needs is compiled in *Appendix F*.

Table 5. Data Gaps Identified in Reviewed HIAs

Data Gaps	
Baseline Data - Data on current health status of communities surrounding site - Baseline health and environmental data - Baseline data on vulnerable populations - Demographics - More detailed baseline health data **Employment Sector/Economic Data** - Psycho-social attributes of jobs (physical work conditions, job security, access to health insurance through employment, lack of control over work, lack of participation in decision-making, time spent at work, work environment, work balance) - Data to estimate impact to customer bills - Wind energy impacts on jobs, income and other economic indicators - Retail effects - Agriculture, forestry, fishing, hunting, mining, quarrying, and oil and gas extraction industry average annual wages - Eligibility requirements for employment and degree to which those positions are fulfilled by residents	**Environmental Data** - Locally-placed air monitors to assess air pollution - Noise assessment - Specific PM2.5 data - Analysis of potential dust/diesel emissions - Visual effects analysis - Off-site terrain noise modeling - Risks of groundwater contamination - Soil testing results indicating whether pesticide chemical residues exist in the soil - Air quality measures other than PM2.5 - Water quality data **Health Data** - Consistently reported health behavior data for Wisconsin youth - Local data on youth alcohol use and effects of underage drinking A study on children and adolescent health

Table 5. Continued

Data Gaps	
Causality Literature/Data - Sufficient research to identify the relative importance of the community design features that promote physical activity - Supplementary research to show causality between elements of the built environment and chronic disease - Parental, social, and environmental factors affecting selection of childcare and school locations (e.g., commute times, etc.) - Local health data (e.g., morbidity/mortality) linked to built environment data - More data on the proximate impacts of markets - Literature that takes socioeconomic status into account when looking at the health impacts of the built environment - Longitudinal studies or randomized controlled trials that further delineate the relationship between the built environment and health - Epidemiological data on the impact of climate change - Health impact data for ozone exposure - More research in the areas of nearby recreation, trail system development and health outcomes in both urban and rural settings - Epidemiological studies on sound, shadow flicker, amplitude modulation, and indoor low frequency sound impacts - Effectiveness of particular interventions for reducing pedestrian injuries - Data related to human consumption of subsistence resources to accurately assess the affects of nutrition changes - Accident rates due to driver distraction **Infrastructure/Services Data** - Data on the quality of public services/infrastructure - Qualitative data on existing bicycle and pedestrian infrastructure - Comprehensive inventory of pedestrian facilities - Additional mapping to fully understand the County's bike network and recreational amenities and their connections to residential areas and other services - Maintenance and repair requirements of transit village - Qualitative data on existing bicycle and pedestrian infrastructure - Comprehensive inventory of pedestrian facilities - Quality of pedestrian environment (PEQI) - Quality of parks and open space - Accessory dwelling unit (ADU) literature for rural areas	**Local/Small-Geographic Scale Data** - City-scale health data (county data used instead) - Vehicle-to-vehicle and vehicle-to-pedestrian accident data - Physical activity data for the quarter-mile radius around the bridge - Identification of pollutant sources in the neighborhood - Mortality and morbidity data by neighborhood - Data on physical activity by neighborhood - Poverty data at a small geographic scale - Overweight/obesity data by zip code - More locale-specific data on the prevalence of pertinent risk factors - Neighborhood-level data on gross per capita water usage and annual per capita waste disposal - Neighborhood-level data on proportion jobs paying self-sufficiency wage and filled by residents, households living on income below self-sufficiency standard, occupational injury, jobs providing health insurance, and proportion locally-owned businesses - Neighborhood-level data on planned parking pricing strategies and traffic calming interventions - Neighborhood-level data on public transit access to public school and proportion children attending neighborhood schools - Neighborhood-level data on access to produce stores and food markets, and neighborhood completeness indicator for key public/retail services - Neighborhood-level data on tree canopy and sidewalks with adequate lighting - Neighborhood-level data on volunteerism - Precise data on displacement - Qualitative data of neighborhood changes to identify communities receiving displaced persons - Income at the block level - Demographic and resources data below zip code level - Record-level local health data (morbidity/mortality) linked to built environment data - Updated subsistence data/analysis **Miscellaneous Data** - Adequate information available to apply the HDMT development target checklist - Data on racial/ethnic disparities (due to small numbers) - Quantitative data to replace qualitative data collected at community/advisory panel meetings - Bike and pedestrian safety data - A more comprehensive study of truck counts and activity - Data on unincorporated areas

Table 5. Continued

Data Gaps	
Permit/Application/Plan Data - More complete/consistent applicant information (data on types, numbers, and age of fleet vehicles; information on type of waste transport and waste transport routes; waste volume, waste origin, and waste characterization; consistent traffic projections) - Quantitative data on number of ADU permits to be requested - Air permit application - Information about park and trail design, entry points, and changes to the surrounding environment to allow a more accurate assessment of access - Details on construction equipment to be used to allow for air pollution and noise impacts during construction to be assessed - More specific implementation strategies - Information to assess compliance with existing housing law - Cost of the project housing - More details to support Sustainable and Safe Transportation and Public Infrastructure and Healthy Economy elements	**Program Data** - Department of Community and Housing Development data on program participants and program utilization - Availability of detailed, accurate information on the expenditures of the School Food Services Branch - Further evaluation of impacts of short-term ODOT solution to safe highway accessibility - Actual impacts of funding cutbacks on transit services - Utility company data on arrearages and shut-offs in Massachusetts **Temporal Data** - Current household level survey data for the potentially affected communities and data for years 2009-2011 - Accident and injury data for years 2009-2011 - More current data than the 2000 and 2006 census 2010 census data for updated population map **Tools** - A mechanism for predicting potential health impacts of proposed land use and policy decisions - Traditional and Local Knowledge survey - Community (citizen) surveys (as study relied on unstructured public comment and literature about community opinions and perceptions) - Estimation of social cohesion and level of physical activity

Baseline Profile

The data gathered in the HIAs from existing sources and new data acquisitions were used, in most cases, to create a profile of existing health conditions and evaluate the potential health impacts of the decision being considered. The profile of existing conditions is used to a) predict future conditions due to impacts of the decision assessed in the HIA and b) compare with future conditions, should the decision be enacted (i.e., for impact monitoring). This baseline profile includes data about the health determinants and outcomes, demographics (e.g., ethnicity, age, gender), and socioeconomic status (e.g., income, poverty, education level, housing value; North American HIA Practice Standards Working Group 2010; National Research Council 2011; Human Impact Partners 2012) of the affected population. Of the 81 HIAs reviewed, 63 created some sort of baseline profile.

Health Endpoints Assessed

As for evaluation of potential health impacts, although the decision was usually made early in the scoping process as to which impacts would be examined, this decision was at times revised as a result of the evidence collected in the assessment step of the HIA process. Table 6 shows the variety of health impacts evaluated in the reviewed HIAs, with the most common health endpoints (i.e., those evaluated in one-quarter or more of the reviewed HIAs) presented in Figure 25.

Table 6. Health Impacts Evaluated in Reviewed HIAs

Health Endpoints		
- attention deficit disorder (ADD)/attention deficit hyperactivity disorder (ADHD) - alcoholism/substance abuse - allergies - anemia - anxiety - arthritis - asthma - behavioral health/development - birth defects - bronchitis - cancer - carbon monoxide poisoning - cardiovascular/circulatory health - central nervous system function - childhood growth/development - cholesterol - chronic disease - chronic obstructive pulmonary disease (COPD) - cognitive function - communicable disease - depression - diabetes - diarrhea - disability - dyslipidemia - emphysema	- endocrine disorders - eye/nose/throat/lung irritation - fatigue - food-borne illness - gallbladder disease - genotoxicity - gynecological/reproductive health - headaches - hearing loss/impairment - heart attack - heart disease - heat/cold related illnesses - hypertension/high blood pressure - immune system/function - infection - infectious disease - inflammation/inflammatory response - injury - irregular heart beat - kidney disease/disorder - lead poisoning - learning disabilities/reduced learning - life expectancy - liver disease/health - low birth weight - lung disease/health	- malnutrition - mental health - metabolic disorder/disease - morbidity - mortality/death/fatality - musculoskeletal/bone & joint - myocardial infarction - nausea - neurological health - nutrition - obesity/weight - osteoporosis - overall/general health - physical health - physiological health - pneumonia - psychological health - rape - respiratory health - sexually transmitted disease - sick building syndrome - sleep apnea - sleep disturbance - stress - stroke - suicide - ulcers - vector borne illness - water borne illness/water toxics exposure

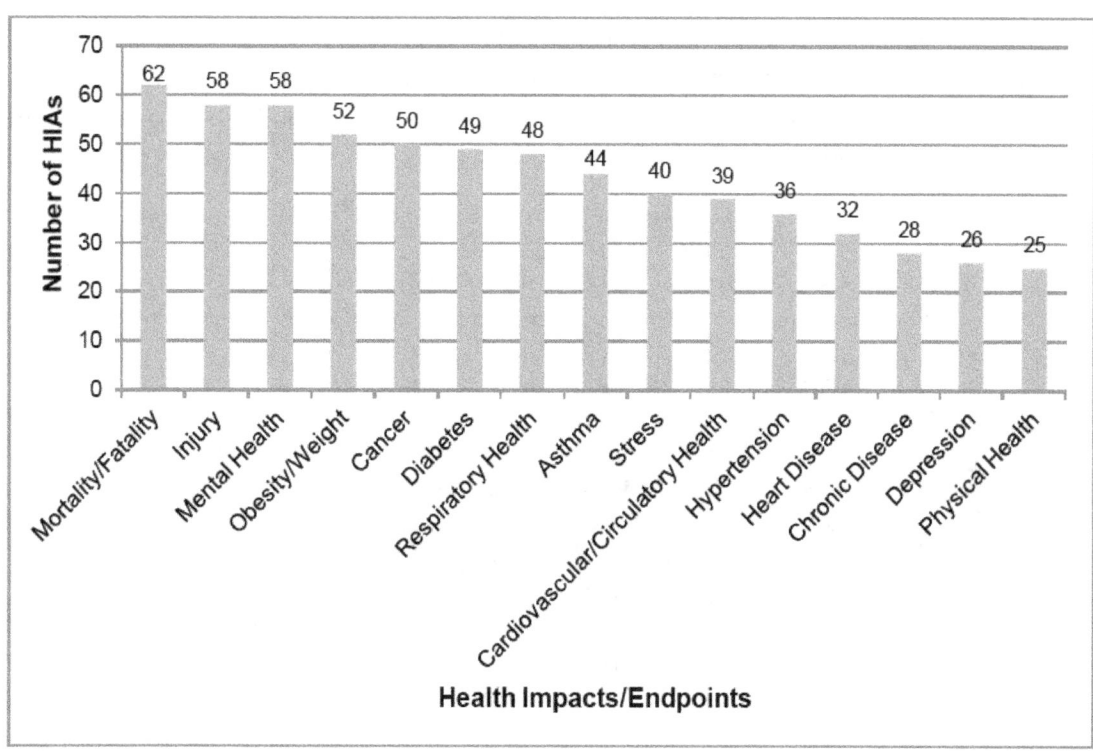

Figure 25. Most common health impacts/endpoints evaluated in reviewed HIAs.

Evidence Defensibility

Predicting the health impacts of a decision with complete certainty is impossible; however, the HIA guidance does call for the use of best available evidence; acknowledgement of assumptions, strengths, and limitations; and transparent synthesis of evidence (North American HIA Practice Standards Working Group 2010; Minimum Elements 2.4, 2.5). Figure 26 shows that not all of the HIAs met these measures for evidence defensibility; however, some HIAs went above and beyond, evaluating and documenting, for instance, the quality of evidence used as the basis for impact assessment (n=9). See *Best Practices* for further discussion on including evaluations of evidence quality in HIAs.

Impact Characterization

In addition to identifying potential health impacts, the assessment step in the HIA process also includes judging the direction, magnitude, likelihood, distribution (i.e., equity), and permanence of impacts via qualitative and quantitative analysis. One trend that became very apparent when reviewing the HIAs was that quantification of impacts was lacking; most HIAs qualitatively characterized impacts. Of the 81 HIAs reviewed, a little more than one-quarter (n=23) employed quantitative analysis in the characterization of impacts. While stakeholder and community input lend themselves to qualitative analysis, many times qualitative analysis was warranted for other reasons as well (e.g., due to lack of available scientific research, unavailability of local data, time limitations, limited resources, etc.) However, there were instances when the use of best available

39

data may have allowed impacts to be quantified. See *Implementation of the HIA Process* for further discussion on the benefits and challenges of quantitative characterization and *Best Practices* for more on the use of best available evidence (both qualitative and quantitative) for health impact characterization.

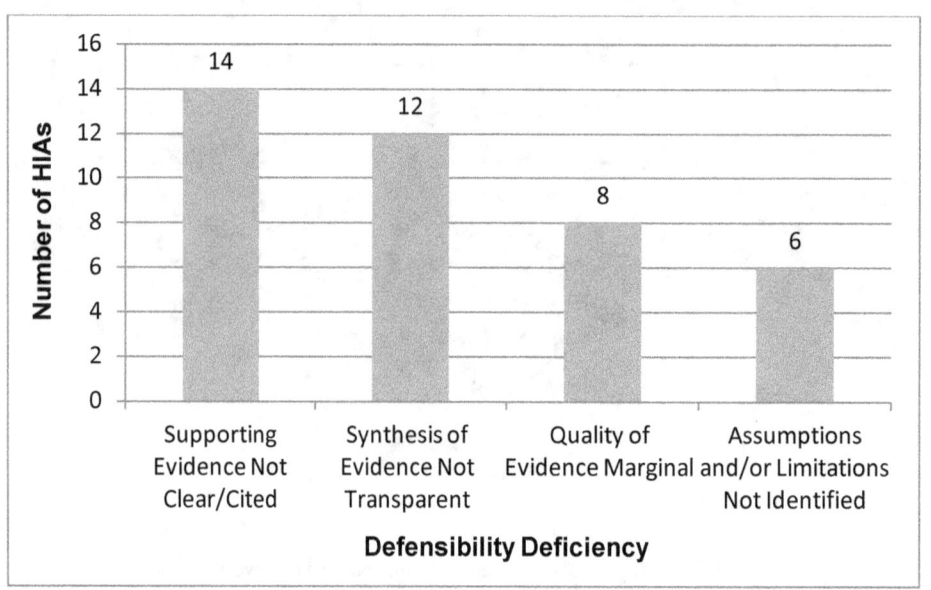

Figure 26. Deficiencies in evidence defensibility in reviewed HIAs.

The characterization of impacts employed by each of the reviewed HIAs is shown in the respective table in *Appendix G*, but Figure 27 provides an overview of how impacts were characterized in the HIAs collectively. As can be seen from the figure, characterization of impacts primarily involved considerations of direction (e.g., positive, negative, unclear, or no effect) and distribution/equity (i.e., identification of disproportionate or equal impacts on the population). Of the 77 HIAs that characterized impacts, 97% included judgement of direction and 88% included judgement of distribution/equity; judgement of likelihood, magnitude, and permanence were each considered in less than half of the reviewed HIAs.

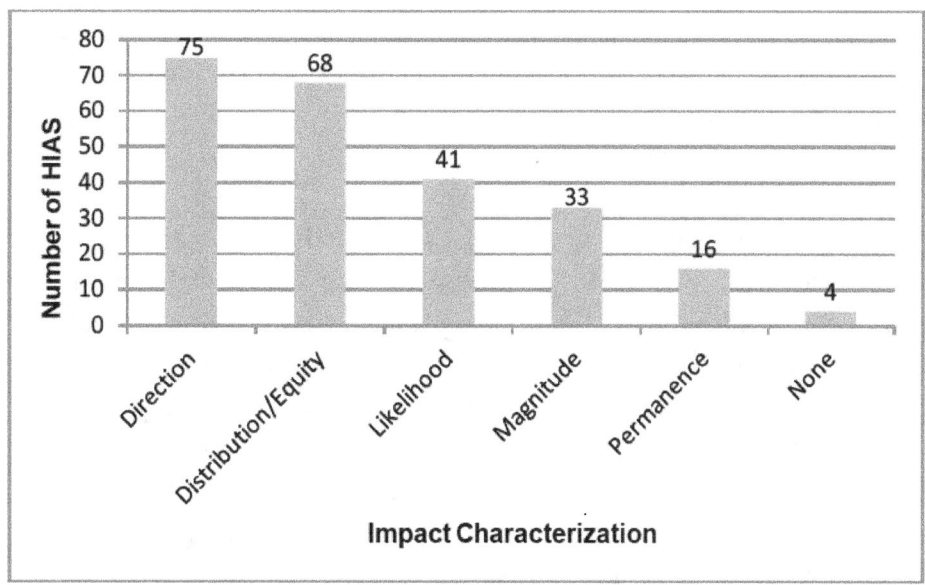

Figure 27. Characterization of impacts in reviewed HIAs.

Recommendations

Once the available data were analyzed and the potential impacts of the decision identified, the next step in the HIA process involved developing recommendations. Recommendations can include support for or opposition to the decision, alternatives or modifications to the decision (e.g., to promote positive health impacts or minimize negative health impacts), or merely mitigations of negative health impacts.

Prioritization of Recommendations

Oftentimes in HIAs, a prioritization process will take place to identify the impacts for which recommendations will be offered and/or identify which of the developed recommendations to offer or prioritize for action. Prioritization can be based on a number of factors, including stakeholder/ community input; distribution/equity of impacts; literature/research; funding availability; cost/economics; impact on health;

> *Recommendations*
>
> Identify strategies for promoting the positive health impacts and/or mitigating the adverse health impacts. Tasks include:
>
> - Developing recommendations (e.g., alternatives to the decision, modifications to the proposed policy/project, mitigation of adverse health impacts)
> - Prioritizing recommendations, if desired
> - Developing an implementation plan for developed recommendations (e.g., responsible party for implementation, timeline, link to indicators that can be monitored)
>
> *Applicable Database Fields*
> - *Prioritization Methods*
> - *Decision-making Outcome*

viability/feasibility of recommendation implementation; quick fixes or initiatives (e.g., that indicate commitment to and lay the foundation for future actions); measurability of impact; quality of evidence; consultation with experts; population affected; or specific prioritization criteria or ranking systems implemented for the project.

Two HIAs implemented unique ranking techniques to classify the health impacts and prioritize recommendations for action. One was a qualitative ranking system that took into consideration four parameters – stakeholder concern, data gaps, potential impact for both positive and negative health effects, and likelihood – to prioritize health effect categories for action by decision-makers, and the other was a four-step risk assessment technique developed by Winkler et al. (2010) that ranks the significance of identified health impacts and prioritizes actions based on the duration, magnitude, extent, likelihood, nature (e.g., direct, indirect, cumulative), and direction of impact. For more details on these unique prioritization approaches, see *Best Practices*. Summarized decision-making outcomes for each HIA, including final recommendations offered in the HIA Report, are identified in the respective table in *Appendix G*.

Implementation Plan or Strategy

In the Recommendations stage of the HIA process, it is also suggested that an implementation plan be developed for the recommendations that includes information such as parties responsible for implementation, timeline, and link to indicators that can be monitored (National Research Council 2011; Human Impact Partners 2012). Implementation plans or strategies for recommendations were found in less than 10% of the HIAs reviewed (n=8).

Reporting

Reporting and communicating the results of the HIA is crucial to informing the decision being evaluated. During the preliminary literature search, there were several instances in which HIA Reports were not publicly accessible, precluding the HIA from being included in the HIA Review. Of the HIAs identified in the final literature search, three were later dropped from the review because the HIA Reports were not available.

Reporting

Write a final report and communicate the results of the HIA to decision-makers for implementation/action. Tasks include:

- Developing a transparent, publicly-accessible HIA Report that documents the process, methods, findings, funding, and participants of the HIA
- Determining the method of communicating HIA findings and recommendations to stakeholders and decision-makers
- Preparing communication materials and communicating the results of the HIA to inform stakeholders and decision-makers

Applicable Database Fields
- *HIA Report*
- *Minimum Elements of HIA Met?*
- *Contact*
- *Organization/HIA Website*

Transparent Documentation of HIA

Not only does the reporting step of the HIA process call for a final report to be prepared that documents the HIA and its results, but the guidelines call for that documentation to be transparent (North American HIA Practice Standards Working Group 2010; National Research Council 2011; Human Impact Partners 2012). Of the 81 HIAs reviewed, over 35% of the HIAs (n=29) lacked transparent documentation of the process, methods, findings, sponsors, funding source(s), and/or participants and their respective roles. Of those 30 HIAs, 22 lacked transparency in funding for the HIA and 6 lacked transparency in identifying an HIA point-of-contact. In contrast, there were a number of HIAs that went above and beyond to ensure that the documentation was transparent, including detailed documentation of assessment methodologies, techniques, and models; criteria for data aggregation; geographic units of analysis (i.e., geographic area and scale); software packages used in analysis; confidence estimates; and supporting documentation, such as screening and scoping worksheets, sources of information used to develop research questions, and methodology, tools, and results of community information gathering (e.g., walkability assessments, interviews, surveys, focus groups, and community and stakeholder meetings).

Communication of HIA Results

In addition to preparing the HIA Report, reporting also involves communicating and disseminating the findings and recommendations of the HIA to inform stakeholders and decision-makers. Only 4 of the 81 HIAs reviewed, made mention of or included in the HIA Report a communication plan or strategy for reporting and disseminating the results of the HIA to the appropriate audiences, although communication plans could have been developed and documented separately from the HIA Report. Methods for communicating the HIA results to decision-makers and stakeholders took many forms in the reviewed HIAs, including dissemination of the HIA Report and/or factsheets, inclusion of HIA documentation in an EIS/EIS public comment period, presentations, press releases, public and/or stakeholder meetings, public testimony in hearings related to the decision, lobbying, Listservs, and personal communication (e.g., letters, emails, phone calls). Secondarily, results of some HIAs were also made available on public websites or published in peer-reviewed journals or regional magazines. For more on the importance of reporting, see *Best Practices* and *Areas for Improvement*.

Monitoring and Evaluation

The monitoring and evaluation step of HIA involves three main forms of evaluation – process evaluation, impact evaluation, and outcome evaluation. Process evaluation involves examining how the HIA process was carried out, including who was involved, strengths and weaknesses of the HIA, successes and challenges, effectiveness in meeting HIA objectives and established practice standards, engagement and communication with stakeholders, and lessons learned. Impact and outcome evaluations are both carried out after completion of the HIA and involve monitoring the impacts of the HIA on the decision and decision-making process (i.e., impact evaluation) and the impacts of the decision implementation on health determinants and outcomes (i.e., outcome evaluation). All three forms of evaluation were lacking in the reviewed HIAs, which unfortunately is not a trend unique to this subset of HIAs (Wismar et al. 2007; National Research Council 2011).

Process Evaluation

Process evaluation, which can be thought of as an evaluation of HIA defensibility and quality, was only found in 5 of the 81 HIAs reviewed, although process evaluations may have been performed separately from these HIAs and not included in the reports.

Monitoring and Evaluation

Evaluate the processes involved in the HIA, the impact of the HIA on the decision-making process, and the impacts of the decision on health. Tasks include:

- Evaluating how the HIA process was carried out, who was involved, how smoothly the assessment proceeded, and how effective the HIA was in meeting its stated objectives and established practice standards (i.e., process evaluation)
- Monitoring decision implementation and the effect the HIA had on the decision-making process (i.e., impact evaluation)
- Monitoring health determinants and outcomes to determine the accuracy of health impact predicted in the HIA (i.e., outcome evaluation), when feasible

Applicable Database Fields
- *Defensibility/Process Evaluation*
- *Effectiveness of HIA*
- *Follow-up Measures*
- *Minimum Elements of HIA Met?*

Impact/Outcome Evaluation

Proposed plans for impact and/or outcome evaluation were present in only 29 of the HIAs. The National Research Council (2011) notes that outcome evaluation is infeasible in many cases, given the length of time between implementation of the decision and changes in health outcomes, as well as the presence of multiple confounding factors contributing to many of the health outcomes. However, in cases when impact and/or outcome evaluation is not feasible, the HIA should discuss the limitations preventing the evaluation from occurring.

In addition to proposals for impact and outcome evaluation, other follow-up and monitoring measures were also identified in the reviewed HIAs. These included securing funding,

44

establishing health baselines and targets for improvement, monitoring specified indicators, performing environmental monitoring (e.g., drinking water supply, emissions, noise, carbon footprint, and air quality monitoring), conducting research to address the information gaps in the HIA, implementing traffic safety monitoring, performing educational outreach, conducting additional HIAs, monitoring complaint systems, monitoring agreements made by developers to ensure they are maintained, disseminating tools developed through the HIA, and modifying existing tools for applicability.

For more on the need for increased monitoring and evaluation in HIA, see *Best Practices* and *Areas for Improvement*.

Assessment of HIA Process – Defensibility, Compliance, and Effectiveness

HIA Defensibility

As part of the HIA Review, reviewers conducted an evaluation of process and defensibility for each HIA. This evaluation took into account the quality of the process undertaken (i.e., evidence and methodology; assumptions, limitations, and barriers; successes and challenges; lessons learned) and the documentation of that process, using the *Minimum Elements and Practice Standards for Health Impact Assessment* (North American HIA Practice Standards Working Group 2010) as a benchmark. Considerations used in the evaluation included, among other things, the questions below.

- Was the supporting information and methodology sound and clearly documented in the report (e.g., adequate literature, data etc. collected; sources of data acknowledged; clear description of data and methodology used; identification of participants and their roles, funding, etc.)?

- Was the scope of the HIA and process undertaken clearly documented?

- Was stakeholder input solicited and utilized?

- Were the recommendations based on transparent, context-specific synthesis of evidence (e.g., impacts/conclusions well supported by the data, literature, etc. presented in the report) or was it not clear how the authors reached the conclusions (e.g., evidence presented only spoke to general health impacts and not the specific impacts examined)?

- Were assumptions, limitations, and barriers present and identified by the authors?

- Was the documentation of the process, methods, findings, sponsors, participants, etc. transparent and publicly-accessible?

HIA Compliance with *Minimum Elements of HIA*

In addition to the evaluation of process and defensibility, reviewers also evaluated each HIA against the *Minimum Elements of HIA* developed by the North American HIA Practice Standards Working Group (2010). These minimum elements are given below.

> "A health impact assessment (HIA) must include the following minimum elements, which together distinguish HIA from other processes. An HIA:
>
> 1. Is initiated to inform a decision-making process, and conducted in advance of a policy, plan, program, or project decision;
>
> 2. Utilizes a systematic analytic process with the following characteristics:
>
>   2.1 Includes a scoping phase that comprehensively considers potential impacts on health outcomes as well as on social, environmental, and economic health determinants, and selects potentially significant issues for impact analysis;
>
>   2.2 Solicits and utilizes input from stakeholders;
>
>   2.3 Establishes baseline conditions for health, describing health outcomes, health determinants, affected populations, and vulnerable sub-populations;
>
>   2.4 Uses the best available evidence to judge the magnitude, likelihood, distribution, and permanence of potential impacts on human health or health determinants;
>
>   2.5 Rests conclusions and recommendations on a transparent and context-specific synthesis of evidence, acknowledging sources of data, methodological assumptions, strengths and limitations of evidence and uncertainties;
>
> 3. Identifies appropriate recommendations, mitigations and/or design alternatives to protect and promote health;
>
> 4. Proposes a monitoring plan for tracking the decision's implementation on health impacts/determinants of concern;
>
> 5. Includes transparent, publicly-accessible documentation of the process, methods, findings, sponsors, funding sources, participants, and their respective roles."

Of the 81 HIAs reviewed, only 13 met <u>all</u> the *Minimum Elements of HIA*. Figure 28 shows the frequency in which each *Minimum Element* was not met (i.e., all of the aspects of the *Minimum Element* were not met or the *Minimum Element* was missing completely).

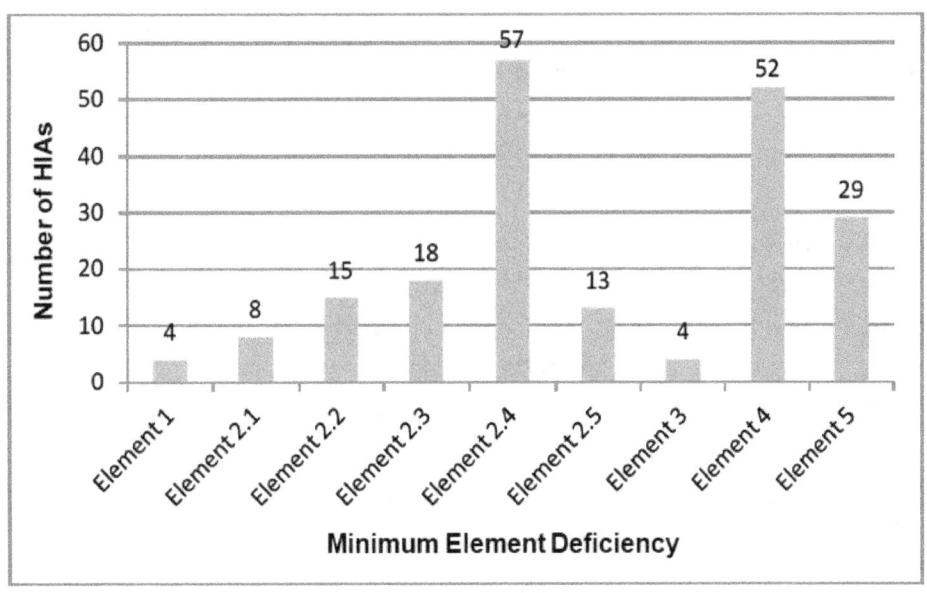

Figure 28. *Minimum Elements* missing or deficient in reviewed HIAs.

HIA Effectiveness

As described previously, impact evaluation involves monitoring the effect of the HIA on the decision-making process and final decision (e.g., how decision-making changed as a result of the HIA, modifications made to the decision as a result of the HIA, and adoption of HIA recommendations and/or mitigations). Because impact evaluation is carried out after completion of the HIA, this measure of effectiveness is not documented as part of the HIA Report. As a result, reviewers subjectively evaluated the effectiveness of each HIA in influencing the outcome of the decision being considered. These evaluations were based on information obtained via internet searches and used four measures of effectiveness as defined by Wismar et al. (2007) – direct effectiveness, general effectiveness, opportunistic effectiveness, and none (i.e., no effectiveness). Direct effectiveness entails the decision being dropped, modified, or postponed as a result of the HIA. General effectiveness, in contrast, involves the HIA being considered by decision-makers, but not resulting in modifications to the proposed decision. One frequent benefit of general effectiveness is often raised awareness of health among decision-makers and stakeholders. In opportunistic effectiveness, the HIA is conducted because it is assumed that the assessment will support the proposed decision; that is, the decision would be carried out regardless of the HIA. In some cases, the HIA is ignored and not taken into account at all by decision-makers; these HIAs have no effectiveness (i.e., none).

Figure 29 shows that it was difficult to discern, via publicly-accessible documentation on the internet, the influence the HIAs had on the decision-making process. While effectiveness could not be discerned for 31 of the reviewed HIAs, of those HIAs for which measures of effectiveness could be obtained (n=50), 60% show direct effectiveness, 32% showed general effectiveness, 6% showed no effectiveness, and 2% showed opportunistic effectiveness. Of the 30 HIAs shown to have a direct effect on the decision, seven required additional testimony or lobbying of decision-makers above and beyond implementation of traditional communication methods.

It should be noted that the measures of effectiveness noted in Figure 29 are subjective and may not reflect the true effect of the HIA on the decision being considered. For instance, the timing of the evaluation of effectiveness may have been such that the full impact of the HIA was not yet realized, or the documentation upon which the measure of effectiveness was based may not have clearly depicted changes in the decision and/or the impetus for those changes.

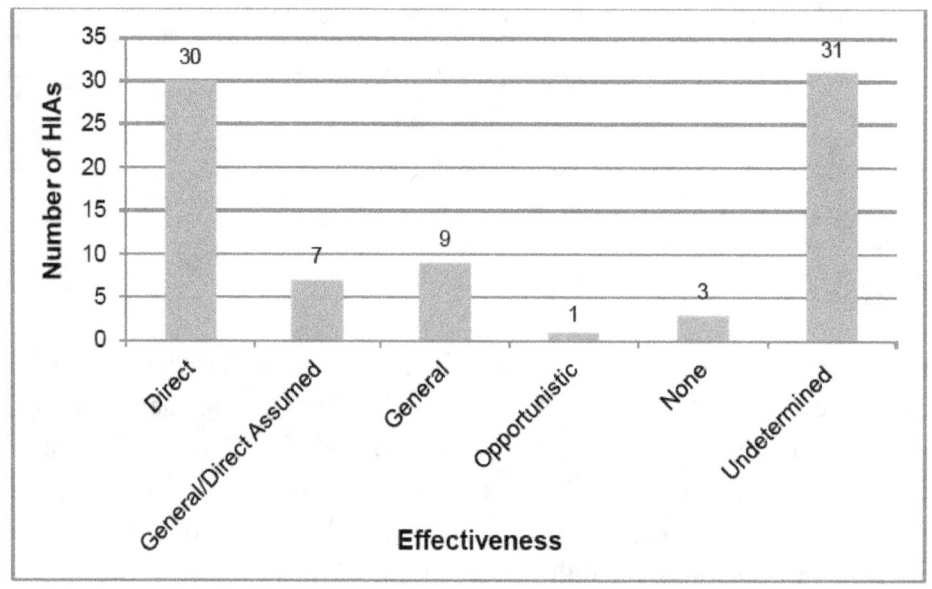

Figure 29. Effect of reviewed HIAs on the decision and decision-making process (based on a subjective evaluation performed by the HIA reviewers).

While 30 of the 81 HIAs reviewed showed direct effectiveness, it is unrealistic to expect that HIAs will influence every decision-making outcome (National Research Council 2011). Decision-makers have to weigh a number of different factors when coming to a decision, and health is but one consideration. Beyond influencing the decision-making outcome, a number of other measures of effectiveness were observed in the reviewed HIAs. These included raised awareness of health and related issues; the introduction of health into discussions where health was typically absent (i.e., informing decision-making); engagement of community members and stakeholders in decisions that affect them; interdepartmental, interagency, and even intersector collaborations; and relationship and capacity building within the community.

While examining the health impacts of implementation of the 81 decisions assessed in the reviewed HIAs was beyond the scope of the HIA Review, during internet searches conducted throughout the review, it became evident that little to no documentation of outcome evaluation is available. This may be due to the lag in time between decision implementation and health outcome changes, the inability to discern whether changes to a specific health outcome is the result of the decision when multiple confounding factors and pathways exist, and the lack of time and resources needed for the long-term research commitment (National Research Council 2011).

Sector Snapshots

This section of the synthesis includes snapshots of the HIAs conducted in each of the four sectors (i.e., Transportation, Housing/Buildings/Infrastructure, Land Use, and Waste Management/Site Revitalization) to identify the types of decisions being assessed in those sectors and the outcomes of the HIAs on those decisions. In addition, each sector snapshot includes a dashboard of summary statistics for the HIAs in that sector; identifies sources of evidence, tools, or methods primary to the sector; and highlights one or two model HIAs from the sector that meet the *Minimum Elements of HIA* and exemplify HIA best practices. Summary tables of select data from the HIA Review Database are provided in *Appendix G* for each of the HIAs.

Transportation Snapshot

The 21 HIAs in the Transportation sector were conducted to assess the impacts of a variety of transportation-related projects, plans, programs, and policies (Table 7). As described previously, the effectiveness of these HIAs in influencing the decisions at hand was evaluated by the HIA reviewers using four measures of effectiveness defined by Wismar et al. (2007) and information obtained via an internet search. As such, the measures of effectiveness noted in Table 7 are subjective and may not reflect the true effect of the HIA on the decision-making process. Likewise, the measures of effectiveness noted in the table, do not necessarily reflect the overall effectiveness of the HIA, but rather the HIA's effect on the decision. For a more detailed discussion of the four measures of effectiveness utilized here and measures of overall HIA effectiveness, see *HIA Effectiveness*.

Table 7. Transportation Decisions Informed by Reviewed HIAs

ID	Title	Decision-making Level	Decision Type	HIA Scope/Summary	HIA Rigor	Effect of HIA on Decision-making
1	Pathways to a Healthy Decatur: A Rapid Health Assessment of the City of Decatur Community Transportation Plan	Local	Plan	Examine the health impacts of the City of Decatur Community Transportation Plan that aims to make Decatur a healthy place to live and work, maintain a high quality of life, and increase opportunities for alternative modes of transportation.	Rapid	General effectiveness assumed at a minimum
5	The Red Line Transit Project Health Impact Assessment	State	Project	Examine current health conditions for the population living in the Red Line corridor, illustrate links between transportation and health in Baltimore, and recommend specific design features and mitigation strategies to maximize the Baltimore Red Line Project's capacity to achieve better health.	Intermediate	General effectiveness at a minimum, although direct effectiveness possible
9	Spokane University District Pedestrian/ Bicycle Bridge Health Impact Assessment	Local	Project	Inform decision makers about potential health impacts that development of a pedestrian bridge in the University District will have on the current and projected population who will live, work, and recreate within a quarter-mile radius of the bridge.	Intermediate	Undetermined
11	The Impact of U.S. Highway 550 Design on Health and Safety in Cuba, New Mexico: A Health Impact Assessment	State	Project	Provide information on how the design of U.S. Highway 550 could impact the health and safety of Cuba area residents and visitors.	Desk-based	General effectiveness at a minimum, although direct effectiveness possible

Table 7. Continued

ID	Title	Decision-making Level	Decision Type	HIA Scope/Summary	HIA Rigor	Effect of HIA on Decision-making
14	Comprehensive Health Impact Assessment: Clark County Bicycle and Pedestrian Master Plan	County	Plan	Examine the likely health impacts of the Clark County Bicycle and Pedestrian Master Plan, whether to adopt the Master Plan or not, and how elements of the Plan could be prioritized to maximize health impacts	Rapid	Direct effectiveness
17	Health Impact Assessment, June 20, 2011: Duluth, Minnesota's Complete Streets Resolution, Mobility in the Hillside Neighborhoods and the Schematic Redesign of Sixth Avenue East	Local	Project	The purpose of the HIA was to determine the potential health impacts of the Sixth Avenue East Schematic Redesign Study, if the redesign was embracing Duluth's Complete Streets Resolution, and how the redesign study could be improved to provide additional health benefits to users of the corridor.	Rapid	Undetermined
27	Mass Transit Health Impact Assessment: Potential Health Impacts of the Governor's Proposed Redirection of California State Transportation Spillover Funds	State	Program/ Policy	Synthesize and communicate research evidence on how proposed cuts in state funding of mass transit may impact the public's health and inform pending transportation funding decisions in California and illustrate how public policies outside the public health and health care sectors can affect public health.	Rapid	General effectiveness

Table 7. Continued

ID	Title	Decision-making Level	Decision Type	HIA Scope/Summary	HIA Rigor	Effect of HIA on Decision-making
32	HIA of the Still/Lyell Freeway Channel in the Excelsior District	Local	Project	Examine the health impacts associated with past construction of the I-280 Freeway and high-traffic surface streets in the Excelsior District of San Francisco after concerns surfaced that residents of that community were being disproportionately exposed to traffic-related exposures, including air pollution, and suffering the health consequences.	Intermediate	Direct effectiveness
42	Columbia River Crossing Health Impact Assessment	State (interstate)	Project	Examine the Columbia River Crossing (CRC) Draft Environmental Impact Statement (EIS) through a public health lens to understand the scope and magnitude of the potential health effects of the four bridge alternatives being considered.	Desk-based	Undetermined
46	The Sellwood Bridge Project: A Health Impact Assessment	County	Project	Assess how the proposed Sellwood Bridge redesign may affect human health during both the construction and operational phases of the project.	Rapid	Undetermined
51	Rapid Health Impact Assessment, Crook County/ City of Prineville, Bicycle and Pedestrian Safety Plan	Local	Plan	Evaluate the current pedestrian and bicycle situation in Prineville, Oregon and provide recommendations to be incorporated into the updated Bicycle and Pedestrian Safety Plan.	Intermediate	Undetermined
53	SR 520 Health Impact Assessment: A Bridge to a Healthier Community	State	Project	Ensure health consequences were considered in the decision-making process for the SR 520 Bridge Replacement and HOV Project and help decision makers evaluate the alternatives based upon their potential health effects.	Intermediate	Opportunistic effectiveness

Table 7. Continued

ID	Title	Decision-making Level	Decision Type	HIA Scope/Summary	HIA Rigor	Effect of HIA on Decision-making
54	A Health Impact Assessment on Policies Reducing Vehicle Miles Traveled in Oregon Metropolitan Areas	State	Program/ Policy	Assess how vehicle miles travelled (VMT) reduction strategies being considered by Oregon's six metropolitan regions would bring about changes in air quality, physical activity, and car accident rates—and what impact that would have on the public's health.	Intermediate	Undetermined
55	Health Effects of Road Pricing In San Francisco, California: Findings from a Health Impact Assessment	Local	Program/ Policy	Examine a future road pricing scenario being studied by the San Francisco County Transportation Authority (SFCTA) that would charge $3 during AM/PM rush hours to travel into or out of the northeast quadrant of San Francisco (which includes a concentration of San Francisco's currently congested downtown streets).	Comprehensive	Undetermined
56	Santa Monica Airport Health Impact Assessment	Local	Project	Organize, analyze, and evaluate existing information and evidence regarding Santa Monica Airport's (SMO's) impact on three issue areas: lack of an airport buffer zone, noise, and air quality.	Rapid	General effectiveness
62	Health Impact Assessment on Transportation Policies in the Eugene Climate and Energy Action Plan	Local	Plan	Examine the positive and negative impacts of transportation policies within the Eugene Climate and Energy Action Plan (CEAP). The HIA examined seven transportation objectives/ recommendations and summarized the scientific evidence that links those policies to health issues in Eugene.	Intermediate	Direct effectiveness

Table 7. Continued

ID	Title	Decision-making Level	Decision Type	HIA Scope/Summary	HIA Rigor	Effect of HIA on Decision-making
65	Health Impact Assessment (HIA) of Proposed "Road Diet" and Restriping Project on Daniel Morgan Avenue in Spartanburg, South Carolina	State	Project	Assess what the expected effect of the proposed Daniel Morgan Avenue (DMA) Road Diet and Restriping Project would be on the safety of motorists, bicyclists, and pedestrians; opportunities for physical activity; opportunities for improved access to goods and services; and air quality.	Intermediate	Undetermined
66	Treasure Island Community Transportation Plan	Local	Plan	Evaluated whether the Treasure Island Transportation Plan met the health needs of its neighborhood residents, using the HDMT assessment tool and focused on ways the transportation system could be designed and implemented to maximize opportunities for active modes of transportation - such as walking and cycling - and minimize the risk of injuries.	Rapid	Direct effectiveness
75	Interstate 75 Focus Area Study Health Impact Assessment	Local	Plan	Review the final recommendations of the Revive Cincinnati: Neighborhoods of the Mill Creek Valley Comprehensive Plan and assess the health impacts of proposed Interstate-75 infrastructure improvements to select neighborhoods adjacent to I-75. Due to time constraints, this HIA only examined health impacts to two of the four focus areas in the study.	Desk-based	Undetermined, but general effectiveness assumed

Table 7. Continued

ID	Title	Decision-making Level	Decision Type	HIA Scope/Summary	HIA Rigor	Effect of HIA on Decision-making
79	Lake Oswego to Portland Transit Project: Health Impact Assessment	Local	Project	Complement the Draft Environmental Impact Statement (DEIS) and more fully assess the health impacts of the three transit alternatives of the Lake Oswego to Portland Transit Project - no-build, enhanced bus service, and streetcar.	Intermediate	Undetermined
81	Health Impact Assessment of the Port of Oakland	Local	Project	Evaluate the cumulative impacts of on-going Port of Oakland growth on the health of residents in West Oakland through multiple inter-related pathways.	Comprehensive	Undetermined

Appendix G, Table G-1 provides a more detailed look at each of these HIAs, including sources of evidence, impacts/endpoints, pathways of impact, characterization of impacts, decision-making outcome, evidence of HIA effectiveness, and evaluation of whether the _Minimum Elements of HIA_ were met. Figure 30 provides a dashboard of summary statistics for the HIAs in this sector and Figure 31 highlights one of the tools utilized in analysis.

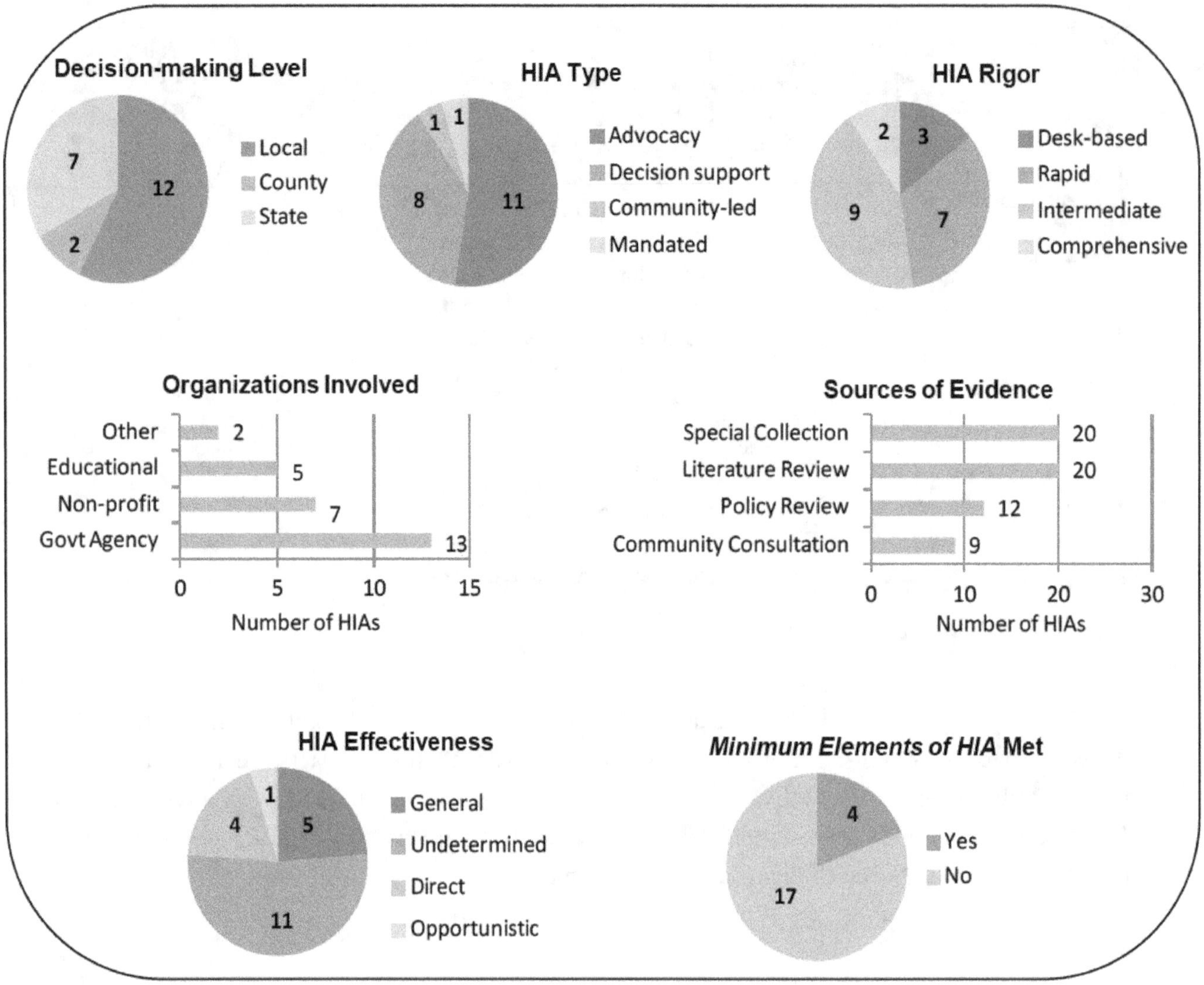

Decision-making Level

Local, County, State

7, 12, 2

HIA Type

Advocacy, Decision support, Community-led, Mandated

1, 1, 8, 11

HIA Rigor

Desk-based, Rapid, Intermediate, Comprehensive

2, 3, 9, 7

Organizations Involved

	Number of HIAs
Other	2
Educational	5
Non-profit	7
Govt Agency	13

Sources of Evidence

	Number of HIAs
Special Collection	20
Literature Review	20
Policy Review	12
Community Consultation	9

HIA Effectiveness

General, Undetermined, Direct, Opportunistic

1, 5, 4, 11

Minimum Elements of HIA Met

Yes, No

4, 17

Figure 30. Dashboard of summary statistics for reviewed HIAs in the Transportation sector.

In the
Spotlight

Health Economic Assessment Tool (HEAT)

HEAT was used in the Health Effects of Road Pricing in San Francisco HIA to calculate the number of lives saved by predicted increases in walking and bicycling under business as usual and the proposed road pricing scenario. Using the EPA value of statistical life, the HIA was able to provide an economic valuation of the mortality averted by active walking and biking. When compared to the economic valuation of the adverse health effects, the HIA found that the cost savings estimated from active transportation via walking and biking were slightly greater than the estimated adverse health costs in each scenario. HEAT is available at: http://www.heatwalkingcycling.org/.

Figure 31. Tool spotlight: Health Economic Assessment Tool (HEAT).

Of the 21 HIAs in the Transportation sector, 15 examined environmental impacts or endpoints (almost exclusively via the air quality pathway), 13 utilized GIS, and 9 utilized modeling (e.g., travel forecasting and traffic noise and emissions modeling). Sources of evidence and data sources primary to the Transportation sector, included expert consultation, data on vehicle miles travelled, and data from the National Transit Database and National Household Travel Survey/ Nationwide Personal Transportation Survey.

Model HIAs from the Transportation sector are shown in Figure 32. These HIAs meet the *Minimum Elements of HIA* and exemplify HIA best practices.

Health Effects of Road Pricing In San Francisco, California: Findings from a Health Impact Assessment

<u>Strengths</u>

- adherence to HIA standards and methodology
- clearly described screening and scoping phases, including factors used in screening
- detailed logic framework (identifying direct, mediating, environmental/behavioral, health and equity impacts; scale of impact; and whether judgement of impact was qualitative or quantitative)
- pathway diagrams showing impacts and data used for each impact type
- sources of evidence and methodology sound and of high quality
- detailed documentation of data sources, geographic units of analysis, and how data was utilized in the HIA
- detailed documentation of assessment methods/models and software packages used
- quantitative evaluation, including economic valuation
- detailed caveats, limitations, and assumptions of assessment
- identified uncertainty factors, assessment approach, and summary confidence level for identified health impacts

Health Impact Assessment (HIA) of Proposed "Road Diet" and Restriping Project on Daniel Morgan Avenue in Spartanburg, South Carolina

<u>Strengths</u>

- adherence to HIA standards and methodology
- identified factors considered in screening
- included scoping worksheet showing research questions, indicators, data sources, and data collection methods in the HIA
- sources of evidence and methodology sound and of high quality
- provided causal pathways
- tabular summary of impacts, including direction, magnitude, likelihood, significance, and distribution

Figure 32. Model HIAs from the Transportation sector.

Housing/Buildings/Infrastructure Snapshot

The 17 HIAs in the Housing/Buildings/Infrastructure sector were conducted to assess the impacts of a variety of Built Environment projects, plans, programs, and policies (Table 8). As described previously, the effectiveness of these HIAs in influencing the decisions at hand was evaluated by the HIA reviewers using four measures of effectiveness defined by Wismar et al. (2007) and information obtained via an internet search. As such, the measures of effectiveness noted in Table 8 are subjective and may not reflect the true effect of the HIA on the decision-making process. Likewise, the measures of effectiveness noted in the table, do not necessarily reflect the overall effectiveness of the HIA, but rather the HIA's effect on the decision. For a more detailed discussion of the four measures of effectiveness utilized here and measures of overall HIA effectiveness, see *HIA Effectiveness*.

Table 8. Housing/Buildings/Infrastructure Decisions Informed by Reviewed HIAs

ID	Title	Decision-making Level	Decision Type	HIA Scope/Summary	HIA Rigor	Effect of HIA on Decision-making
2	Affordable Housing and Child Health: A Child Health Impact Assessment of the Massachusetts Rental Voucher Program	State	Program/ Policy	Evaluate the implications of the Massachusetts Rental Voucher Program (MRVP), a housing assistance and homelessness prevention program, and proposed MRVP changes for FY2006, for children's health and well-being.	Intermediate	Direct effectiveness
8	Health Impact Assessment: South Lincoln Homes, Denver CO	Local	Plan	Examine the redevelopment master plan for the Denver Housing Authority's South Lincoln Homes community in Downtown Denver for potential impacts the redevelopment may have on health and wellbeing of the South Lincoln neighborhood.	Comprehensive	Direct effectiveness
12	Community Health Assessment: Bernal Heights Preschool - An Application of the Healthy Development Measurement Tool (HDMT)	Local	Project	Inform decision making processes related to the choice among three potential future locations of the Bernal Heights Preschool.	Rapid	Undetermined

58

Table 8. Continued

ID	Title	Decision-making Level	Decision Type	HIA Scope/Summary	HIA Rigor	Effect of HIA on Decision-making
23	Health Impact Assessment of Modifications to the Trenton Farmer's Market (Trenton, New Jersey)	Local	Project	Examine several proposed changes to a farmers market in Trenton, New Jersey, including two being considered by the market's executive board (i.e., minor cosmetic changes and a major remodel) and a third suggestion (i.e., a market outreach strategy) and their impacts on patrons' nutrition and physical activity patterns, as well as the potential economic and social capital benefits for vendors and the surrounding community.	Intermediate	Undetermined, but no effectiveness assumed
28	The Rental Assistance Demonstration Project - A Health Impact Assessment	Federal	Program/ Policy	Examine the impacts of a proposed federal housing policy designed to address some of the systemic funding issues related to public housing on a number of health determinants that remained unanswered in legislative debates; and ensure that the evaluation of this pilot project comprehensively considered the health impacts of public housing-related policy decisions	Comprehensive	General effectiveness
30	Jack London Gateway Rapid Health Impact Assessment: A Case Study	Local	Project	Examine a planned retail expansion and low-income senior housing development and address community concerns about air quality, noise, safety, and retail planning	Rapid	Direct effectiveness
35	A Health Impact Assessment of Accessory Dwelling Unit Policies in Rural Benton County, Oregon	County	Program/ Policy	Examine the impacts of five accessory dwelling unit policy options, ranging from restricting currently permitted uses to allowing construction of a complete accessory unit.	Comprehensive	Direct effectiveness

Table 8. Continued

ID	Title	Decision-making Level	Decision Type	HIA Scope/Summary	HIA Rigor	Effect of HIA on Decision-making
36	The Health Impact Assessment (HIA) of the Commonwealth Edison (ComEd) Advanced Metering Infrastructure (AMI) Deployment	State	Program/ Policy	Identify the impact of advanced metering infrastructure (AMI) deployment on the health of residential customers in the Commonwealth Edison (ComEd) service territory in Illinois, particularly vulnerable customers - the very young (birth to age 5), older individuals (age 65+), individuals with functional disability status including those with temperature sensitive conditions, individuals who are socially isolated, and individuals with limited English proficiency or literacy.	Comprehensive	Direct effectiveness
40	Unhealthy Consequences: Energy Costs and Child Health - A Child Health Impact Assessment of Energy Costs and the Low Income Home Energy Assistance Program	State	Program/ Policy	Evaluate impacts of both home heating and total home energy (including electricity, water heating, and cooking) costs and a federally-funded energy assistance program - the Low Income Home Energy Assistance Program (LIHEAP) -on the health of children.	Intermediate	Direct effectiveness
48	A Rapid Health Impact Assessment of the City of Los Angeles' Proposed University of Southern California Specific Plan	Local	Plan	Examine how the proposed University of Southern California (USC) Specific Plan would impact measures of housing, gentrification, and displacement and lead to changes in health for the communities around the USC campus, particularly low-income and vulnerable populations.	Rapid	Direct effectiveness

Table 8. Continued

ID	Title	Decision-making Level	Decision Type	HIA Scope/Summary	HIA Rigor	Effect of HIA on Decision-making
50	Anticipated Effects of Residential Displacement on Health: Results from Qualitative Research	Local	Project	Examine the Trinity Plaza Redevelopment, which proposed to demolish an older apartment building with over 360 rent-controlled units and replace it with 1,400 market-rate condominiums and the potential effects of eviction on health and well-being of tenants.	Rapid	Direct effectiveness
52	29th St. / San Pedro St. Area Health Impact Assessment	Local	Project	Ensure that health impacts were considered in the development plan for The Crossings at 29th Street - an 11.6-acre development that included affordable housing and retail and community space - and in the broader policies impacting redevelopment in the area.	Intermediate	Direct effectiveness
57	Lowry Corridor, Phase 2 Health Impact Assessment	Local	Project	Analyze the potential health effects of Phase 2 development of the Lowry Avenue Corridor Project, a five-mile thoroughfare located north of downtown Minneapolis.	Desk-based	Direct effectiveness
61	Hospitals and Community Health HIA: A Study of Localized Health Impacts of Hospitals	Local	Project	Built upon the Atlanta BeltLine HIA to retrospectively examine the localized health impacts of Piedmont Hospital - one of the major anchor institutions along the Peachtree Corridor in Atlanta, Georgia - and prospectively examine how plans for future growth could change those impacts.	Comprehensive	General effectiveness

Table 8. Continued

ID	Title	Decision-making Level	Decision Type	HIA Scope/Summary	HIA Rigor	Effect of HIA on Decision-making
70	Pathways to Community Health: Evaluating the Healthfulness of Affordable Housing Opportunity Sites Along the San Pablo Avenue Corridor Using Health Impact Assessment	Local	Plan	Assess the health impacts associated with the San Pablo Avenue Specific Plan for three sites proposed to be included in a campaign for affordable housing, and encourage the healthfulness of the San Pablo Area Specific Plan and eventual site development	Intermediate	Undetermined, but general effectiveness assumed
76	A Rapid Health Impact Assessment of the Long Beach Downtown Plan	Local	Plan	Ensure decisions in the City of Long Beach Downtown Plan and Long Beach Downtown Plan Environmental Impact Report account for impacts to low-income and vulnerable populations in the areas of housing and employment.	Rapid	None
80	HOPE VI to HOPE SF San Francisco Public Housing Redevelopment: A Health Impact Assessment	Local	Program/ Policy	Explore the positive and negative health impacts of past Housing Opportunities for People Everywhere (HOPE) VI redevelopment at two sites - Bernal Dwellings and North Beach Place - with the aim of finding opportunities to address existing problems and informing future public housing redevelopment in the HOPE SF Program.	Intermediate	General effectiveness at a minimum, but direct effectiveness assumed

Appendix G, Table G-2 provides a more detailed look at each of these HIAs, including sources of evidence, impacts/endpoints, pathways of impact, characterization of impacts, decision-making outcome, evidence of HIA effectiveness, and evaluation of whether the *Minimum Elements of HIA* were met. Figure 33 provides a dashboard of summary statistics for the HIAs in this sector and Figure 34 highlights one of the tools utilized in analysis.

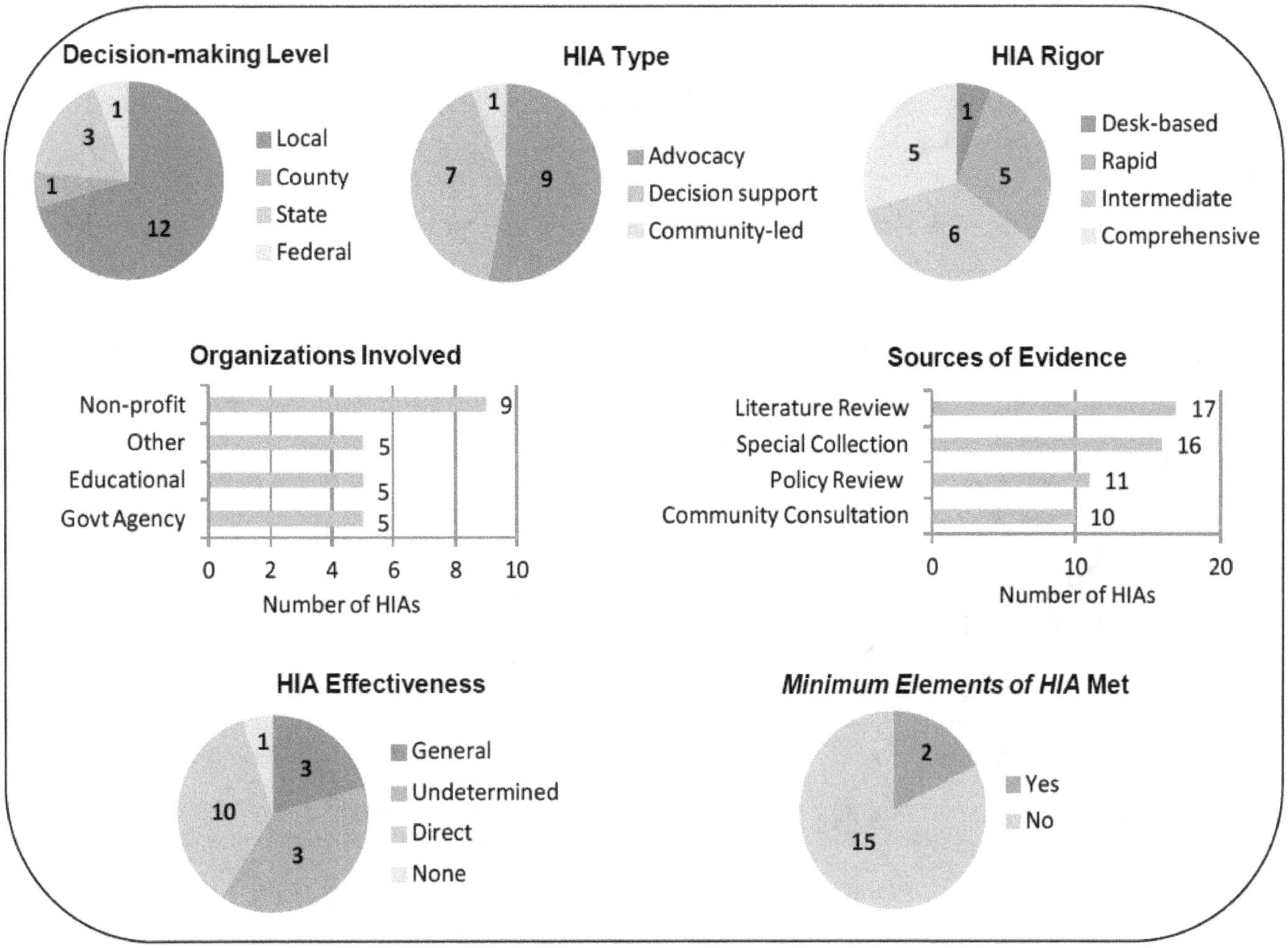

Decision-making Level

- Local
- County
- State
- Federal

(12, 1, 3, 1)

HIA Type

- Advocacy
- Decision support
- Community-led

(9, 7, 1)

HIA Rigor

- Desk-based
- Rapid
- Intermediate
- Comprehensive

(5, 5, 6, 1)

Organizations Involved

	Number of HIAs
Non-profit	9
Other	5
Educational	5
Govt Agency	5

Sources of Evidence

	Number of HIAs
Literature Review	17
Special Collection	16
Policy Review	11
Community Consultation	10

HIA Effectiveness

- General
- Undetermined
- Direct
- None

(3, 3, 10, 1)

Minimum Elements of HIA Met

- Yes
- No

(2, 15)

Figure 33. Dashboard of summary statistics for reviewed HIAs in the Housing/Buildings/Infrastructure sector.

In the Spotlight

Traffic Noise Model (TNM) and Noise Annoyance Relationship

The TNM was used to estimate traffic noise levels at two potential affordable housing development sites in the HIA Evaluating the Healthfulness of Affordable Housing Opportunity Sites Along the San Pablo Avenue Corridor. This data was used in combination with the Miedema and Oudshoorn Noise Annoyance Relationship to estimate the percentage of population that would be highly annoyed by the road traffic noise at these two sites. The TNM is available at: http://www.fhwa.dot.gov/environment/noise/traffic_noise_model/; and the Noise Annoyance Relationship is available at: http://www.ncbi.nlm.nih.gov/pmc/articles/PMC1240282/pdf/ehp0109-000409.pdf

Figure 34. Tool spotlight: Traffic Noise Model (TNM) and Noise Annoyance Relationship.

Of the 17 HIAs in the Housing/Buildings/Infrastructure sector, 8 were conducted in California and 1 was a nationwide HIA; five (5) HIAs examined environmental impacts or endpoints, 10 utilized GIS, and 4 utilized modeling (e.g., traffic noise and emissions modeling). Data sources and tools primary to the Housing/Buildings/Infrastructure sector included HUD, the American Housing Survey, and the American Community Survey.

A model HIA from the Housing/Buildings/Infrastructure sector is shown in Figure 35. This HIA meets the *Minimum Elements of HIA* and exemplifies HIA best practices.

The Rental Assistance Demonstration Project - A Health Impact Assessment

<u>Strengths</u>
- adherence to HIA standards and methodology
- documentation of each step of the HIA process, including screening
- included scoping worksheet, housing/health survey questions, and focus group moderator's guide in the HIA
- sources of evidence and methodology sound and of high quality
- used case study cities to focus and ground findings for nationwide HIA
- included a section in the HIA Report identifying the most challenging limitations
- provided pathway diagrams, research questions, empirical analysis, predicted impacts, and recommendations for each determinant of health
- tabular summary identifying direction, magnitude, and severity of impacts, evidence strength, and uncertainties for each health determinant

Figure 35. Model HIA from the Housing/Buildings/Infrastructure sector.

Land Use Snapshot

The 39 HIAs in the Land Use sector were conducted to assess the impacts of a variety of land use-related projects, plans, programs, and policies (Table 9). As described previously, the effectiveness of these HIAs in influencing the decisions at hand was evaluated by the HIA reviewers using four measures of effectiveness defined by Wismar et al. (2007) and information obtained via an internet search. As such, the measures of effectiveness noted in Table 9 are subjective and may not reflect the true effect of the HIA on the decision-making process. Likewise, the measures of effectiveness noted in the table, do not necessarily reflect the overall effectiveness of the HIA, but rather the HIA's effect on the decision. For a more detailed discussion of the four measures of effectiveness utilized here and measures of overall HIA effectiveness, see *HIA Effectiveness*.

Table 9. Land Use Decisions Informed by Reviewed HIAs

ID	Title	Decision-making Level	Decision Type	HIA Scope/Summary	HIA Rigor	Effect of HIA on Decision-making
3	Health Impact Assessment - Derby Redevelopment, Historic Commerce City, Colorado	Local	Project	Evaluate potential impact of Derby's redevelopment on physical activity and nutrition behaviors of the population of historic Commerce City.	Rapid	General effectiveness assumed at a minimum
6	Health Impact Assessment Report: Alcohol Environment - Village of Weston, WI	Local	Program/ Policy	Assess the impact of an alcohol policy on the community's health, specifically underage drinking and drinking and driving behaviors. While there was no specific policy under review at the onset of the project, the potential impacts of a retail outlet density policy, specifically a limit on future Class A alcohol licenses, on community health and development were assessed.	Intermediate	Undetermined
7	Eastern Neighborhoods Community Health Impact Assessment Final Report	Local	Project	Assess the health benefits and burdens of development, land use plans, and zoning controls in several San Francisco neighborhoods, including the Mission, South of Market, and Portero Hill.	Comprehensive	General effectiveness at a minimum, but direct effectiveness is assumed

Table 9. Continued

ID	Title	Decision-making Level	Decision Type	HIA Scope/Summary	HIA Rigor	Effect of HIA on Decision-making
10	Health Impact Assessment: An Analysis of Potential Sites for a Regional Recreation Center to Serve North Aurora, Colorado	Local	Program/ Policy	Inform a policy decision about the specific location of a regional recreation center in North Aurora, identify impacts to health, and provide recommendations for the Aurora Residents for Recreation Task Force (ARRTF), City Planners, and City Council.	Rapid	Undetermined
13	St. Louis Park Comprehensive Plan - Health Impact Assessment	Local	Plan	Assess the St. Louis Park Comprehensive Plan to ensure that public health is considered within the plan.	Desk-based	General effectiveness at a minimum, but possibly direct effectiveness
15	Health Impact Assessment: Key Recommendations of the Northeast Area Plan	Local	Plan	Evaluate the six key recommendations of the City of Columbus Northeast Area Plan with respect to physical activity for the residents of the Northeast area.	Desk-based	Undetermined
16	Yellowstone County/City of Billings Growth Policy Health Impact Assessment	Local	Program/ Policy	Take a retrospective look at the Growth Policy that was adopted in 2003 in order to identify ways to make health a part of the decision making process regarding community growth by predicting health consequences, informing decision makers and the public about health impacts, and providing realistic recommendations to prevent or mitigate negative health outcomes.	Intermediate	Direct effectiveness

Table 9. Continued

ID	Title	Decision-making Level	Decision Type	HIA Scope/Summary	HIA Rigor	Effect of HIA on Decision-making
18	Knox County Health Department Community Garden Health Impact Assessment: Recommendations for Lonsdale, Inskip and Mascot	County	Program/Policy	Inform policy decisions related to the placement and maintenance of community gardens in Knox County, Tennessee and to objectively present the facts surrounding community gardens and why zoning code should be changed if needed in order to support their placement within residential and nonresidential communities	Desk-based	Direct effectiveness
19	Alaska Outer Continental Shelf - Beaufort Sea and Chukchi Sea Planning Areas, Oil and Gas Lease Sales 209, 212, 217, and 221 Draft Environmental Impact Statement; Appendix J - Public Health	Federal	Project	Examine the health impacts of the proposals for oil and gas leasing in the Beaufort and Chukchi seas, as well as the 10 alternatives to these proposed actions addressed in the EIS.	Intermediate	General effectiveness assumed
20	Divine Mercy Development Health Impact Assessment	State	Plan	Inform recommendations on incorporating health and climate change indicators into the Minnesota Environmental Assessment Worksheet (EAW) used in the environmental review process.	Desk-based	Undetermined
21	Fort McPherson Rapid Health Impact Assessment: Zoning for Health Benefit to Surrounding Communities During Interim Use	Local	Project	As part of a project to bring a Health in all Policies (HiAP) perspective into the baseline realignment and closure process for Fort McPherson and assessed the zoning provisions that govern permitted uses of land, green space, and transportation to gauge their effect on health.	Rapid	General effectiveness at a minimum, but possibly direct effectiveness

Table 9. Continued

ID	Title	Decision-making Level	Decision Type	HIA Scope/Summary	HIA Rigor	Effect of HIA on Decision-making
22	Re: November 10th Merced County General Plan Update (MCGPU) Preferred Growth Alternative Decision	County	Plan	Examine the two growth alternatives being considered for the Merced County General Plan Update – one that focused development in existing urban areas and another that would allow for the creation of new towns in the county – and associated health impacts.	Desk-based	None
24	SE 122nd Avenue Planning Study Health Impact Assessment	Local	Project	Evaluate both the health impacts of the SE 122nd Avenue Pilot Project recommendations themselves, as well as the health impacts of the 20-minute neighborhood form.	Intermediate	Undetermined, but general effectiveness assumed at a minimum
26	Health Impact Assessment: Hawai'i County Agriculture Development Plan	County	Plan	Evaluate the potential positive and negative impacts of three Agriculture Plan policies - institutional buying (farm-to-school programs), commercial expansion of food agriculture, and home production -on the health of Hawaii Island residents.	Intermediate	Direct effectiveness
29	Case Study: Bloomington Xcel Energy Corridor Trail Health Impact Assessment	Local	Project	Assess potential health impacts and obstacles to the proposed Xcel recreational trail corridor and support for including the Xcel trail corridor in the Alternative Transportation Plan	Rapid	General effectiveness
31	Health Impact Assessment for Proposed Coal Mine at Wishbone Hill, Matanuska-Susitna Borough Alaska	State	Project	Review potential positive and negative human health impacts related to the proposed Wishbone Hill Mine (WHM) - a surface coal mine located in the Matanuska-Susitna valley near Sutton, Alaska.	Intermediate	Undetermined

Table 9. Continued

ID	Title	Decision-making Level	Decision Type	HIA Scope/Summary	HIA Rigor	Effect of HIA on Decision-making
34	City of Ramsey Health Impact Assessment	Local	Plan	Assess the potential health impacts of current city planning practices, set goals for improvement, and develop future policy directions in conjunction with the 2008 City of Ramsey Comprehensive Plan update.	Desk-based	Direct effectiveness
37	Atlanta Beltline Health Impact Assessment	Local	Project	Make health a part of the Atlanta BeltLine decision-making process by predicting health consequences, informing decision makers and the public about health impacts, and providing realistic recommendations to prevent or mitigate negative health outcomes.	Comprehensive	Undetermined
38	Zoning for a Healthy Baltimore: A Health Impact Assessment of the Transform Baltimore Zoning Code Rewrite	Local	Project	Evaluate the impacts of Baltimore's comprehensive zoning code rewrite, TransForm Baltimore, to maximize the potential for the zoning recode to prevent obesity and other adverse health outcomes and reduce inequities in these outcomes among children and adolescents in Baltimore.	Intermediate	Undetermined
39	Hood River County Health Department Health Impact Assessment for the Barrett Property	County	Project	Investigate the potential health benefits of turning a former orchard into a community park with open play fields, trails, and community gardens and the potential health risks for users of the property from exposure to residual pesticide chemicals.	Intermediate	Undetermined

Table 9. Continued

ID	Title	Decision-making Level	Decision Type	HIA Scope/Summary	HIA Rigor	Effect of HIA on Decision-making
41	Technical Report 9: Highway 99 Sub-Area Plan Health Impact Assessment	County	Plan	Support the Sub-Area Plan vision (to apply land use planning to build a healthy community) by using an established socio-ecological model of health promotion to validate the plan's health promoting features.	Rapid	Direct effectiveness
43	Inupiat Health and Proposed Alaskan Oil Development: Results of the First Integrated Health Impact Assessment/ Environmental Impact Statement for Proposed Oil Development on Alaska's North Slope	Federal	Project	Developed as part of a supplemental Environmental Impact Statement (EIS) to examine health impacts of oil and gas development in the Teshekpuk Lake Special Area of the Northeast National Petroleum Reserve (NPR)-A (North Slope Bureau, Alaska).	Intermediate	Direct effectiveness
44	Page Avenue Health Impact Assessment	Local	Project	Provide an impartial assessment of the health impacts of the Page Avenue Redevelopment on individuals, youth, and families living primarily in Pagedale, Missouri as well as surrounding communities in University City and Wellston.	Comprehensive	Undetermined
45	Pittsburg Railroad Avenue Specific Plan Health Impact Assessment	Local	Plan	Determine the health impacts of the Pittsburg Railroad Avenue Specific Plan - a transit-oriented design plan to build a new train station, new residential and commercial uses, public space, and pedestrian and bicycle improvements.	Comprehensive	Direct effectiveness

Table 9. Continued

ID	Title	Decision-making Level	Decision Type	HIA Scope/Summary	HIA Rigor	Effect of HIA on Decision-making
47	The East Bay Greenway Health Impact Assessment	Unknown	Project	Highlight potential positive impacts of the Greenway pedestrian and bike trail could have on health and to uncover and suggest mitigations for potential barriers that would hinder the project from reaching its full positive health impact.	Rapid	Direct effectiveness
49	Taylor Energy Center Health Impact Assessment	County	Project	Analyze the impact of a proposed coal-fired electric plant, including risks from air pollution and benefits to health from employment by the plant and the "community contribution"	Rapid	Undetermined
58	Battlement Mesa Health Impact Assessment (2nd Draft)	County	Project	Address citizen concerns about health impacts of natural gas development and production in the Battlement Mesa Planned Unit Development (PUD).	Comprehensive	Direct effectiveness
59	Douglas County Comprehensive Plan Update Health Impact Assessment	County	Plan	Evaluate updates to the Douglas County Comprehensive Plan, which provides a framework and policy direction for future land use, transportation, natural resource and park/open space decisions.	Intermediate	Direct effectiveness
60	The Executive Park Subarea Plan Health Impact Assessment: An Application of the Healthy Development Measurement Tool (HDMT)	Local	Plan	Summarize the results from the first application of San Francisco's Healthy Development Measurement Tool to the Executive Park Subarea Plan, which proposes to build 2,800 units of new residential housing on a 71-acre area in the southeastern corner of San Francisco.	Comprehensive	Undetermined
63	Oak to Ninth Avenue Health Impact Assessment	Local	Project	Assess the influence of the Oak to Ninth Avenue development project – a waterfront mixed-use neighborhood - on determinants of human health.	Intermediate	Direct effectiveness

Table 9. Continued

ID	Title	Decision-making Level	Decision Type	HIA Scope/Summary	HIA Rigor	Effect of HIA on Decision-making
64	A Health Assessment of Mixed Use Redevelopment Nodes and Corridors in Lincoln, Nebraska	County	Plan	Analyze the nodes and corridors proposal in the Comprehensive Plan to determine whether the proposed changes would truly generate health benefits in Lincoln.	Rapid	None
67	Healthy Tumalo Community Plan: A Health Impact Assessment on the Tumalo Community Plan; A Chapter Of The 20-Year Deschutes County Comprehensive Plan Update	County	Plan	Evaluate the draft Tumalo Community Plan in the context of community health and support county planners by providing recommendations that could be incorporated into the final plan.	Intermediate	Direct effectiveness
68	Strategic Health Impact Assessment on Wind Energy Development in Oregon (Public Review Draft)	State	Project	Assess ways that wind energy developments in Oregon might affect the health of individuals and communities where they are built and maintained, develop evidence-based recommendations for future facility siting decisions, engage community and stakeholders, and assess the utility of HIA for specific wind farm siting decisions.	Comprehensive	Undetermined
69	Impacts on Community Health of Area Plans for the Mission, East SoMa, and Potrero Hill/Showplace Square: An Application of the Healthy Development Measurement Tool	Local	Plan	Use the Healthy Development Measurement Tool (HDMT) to examine potential health implications of the Eastern Neighborhoods Area Plans, using 26 of 27 community health objectives within six healthy city vision elements - environmental steward-ship, sustainable and safe transportation, social cohesion, public infrastructure/access to goods and services, adequate and healthy housing, and healthy economy.	Comprehensive	Direct effectiveness

Table 9. Continued

ID	Title	Decision-making Level	Decision Type	HIA Scope/Summary	HIA Rigor	Effect of HIA on Decision-making
71	MacArthur BART Transit Village Health Impact Assessment	Local	Project	Examine the health impacts of the MacArthur BART Transit Village - a proposed redevelopment of the MacArthur Bay Area Rapid Transit Station parking lot and adjacent property into a mixed use village.	Comprehensive	Undetermined
72	Healthy Corridor for All: A Community Health Impact Assessment of Transit-oriented Development Policy in St. Paul Minnesota	Local	Program/ Policy	Examine the rezoning ordinance that would lay the foundation for the implementation of transit-oriented development (TOD) along the Central Corridor to understand the impacts of the light rail line and subsequent land use changes on community health, health inequities, and underlying conditions that determine health.	Comprehensive	General effectiveness at a minimum, but direct effectiveness assumed
73	Health Impact Assessment Point Thomson Project	Federal	Project	Identify human health impacts associated with each of the five proposed design alternatives of the proposed oil and gas development in Alaska's remote Point Thomson area.	Intermediate	General effectiveness
77	Humboldt County General Plan Update Health Impact Assessment	County	Plan	Evaluate six key areas of the Humboldt County General Plan Update (GPU) to identify how indicators of healthy development would change as a result of the three alternatives being considered - denser development in urban areas, limited growth to exurban areas, and unrestricted growth across the county.	Intermediate	Direct effectiveness
78	Rapid Health Impact Assessment: Vancouver Comprehensive Growth Management Plan 2011	Local	Plan	Examine the 2011 Vancouver Comprehensive Plan and its impact on two key determinants of health - physical activity and access to healthy food.	Rapid	Undetermined

Appendix G, Table G-3 provides a more detailed look at each of these HIAs, including sources of evidence, impacts/endpoints, pathways of impact, characterization of impacts, decision-making outcome, evidence of HIA effectiveness, and evaluation of whether the *Minimum Elements of HIA* were met. Figure 36 provides a dashboard of summary statistics for the HIAs in this sector and Figure 37 highlights one of the tools utilized in analysis.

Of the 39 HIAs in the Land Use sector, 9 were conducted in California and 6 were conducted in Minnesota; twenty-six (26) HIAs examined environmental impacts or endpoints, 28 utilized GIS, and 15 utilized modeling (e.g., travel forecasting and urban emission, air pollutant dispersion, economic, food availability, and parking demand modeling). Data sources and tools primary to the Land Use sector, included zoning data, data from the Bureau of Labor Statistics, and the CALINE3/CAL3QHC/CAL3QHCR models (for predicting pollutant dispersion from traffic).

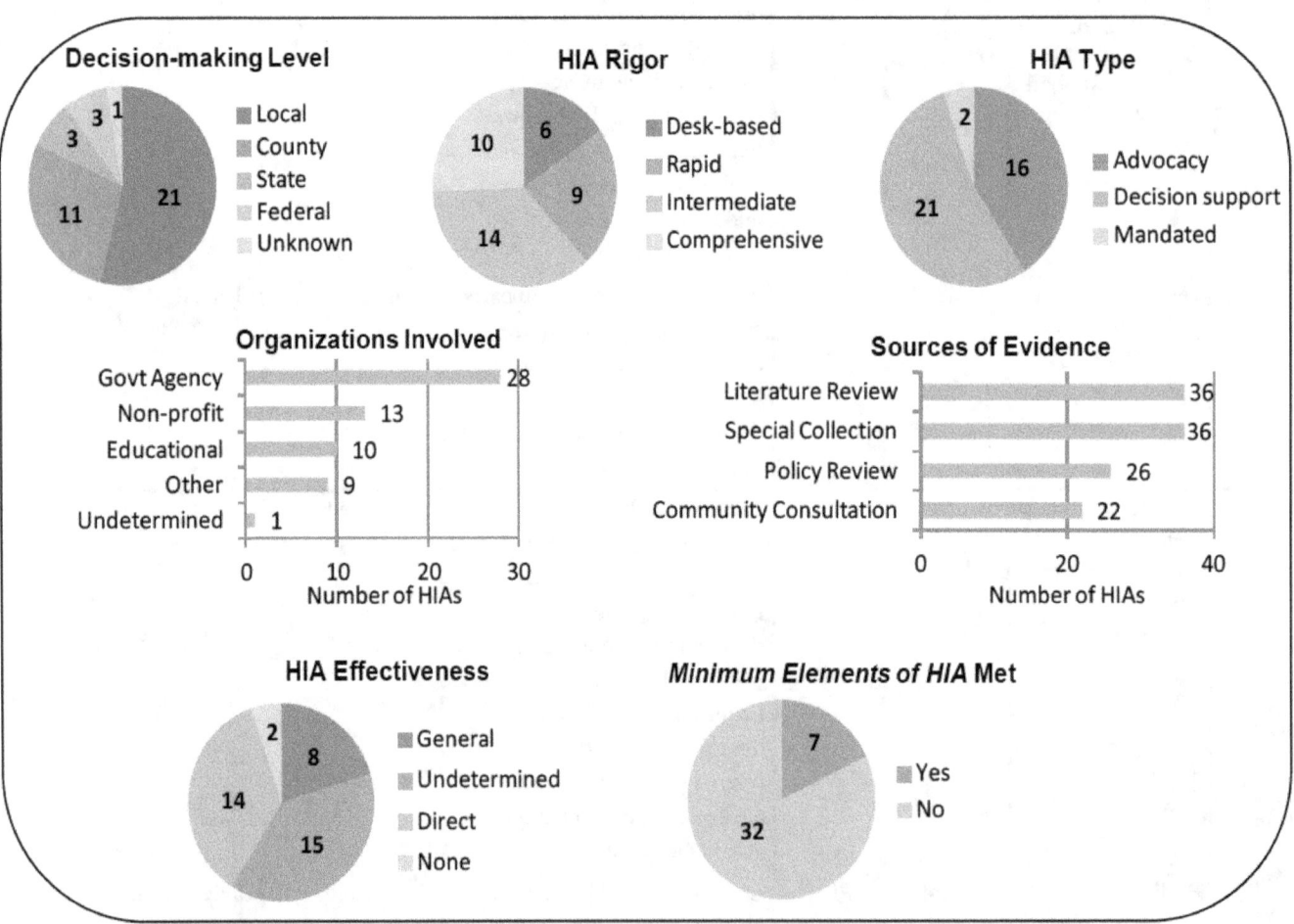

Figure 36. Dashboard of summary statistics for reviewed HIAs in the Land Use sector.

In the Spotlight

Figure 37. Tool spotlight: Walk Score.

Model HIAs from the Land Use sector are shown in Figure 38. These HIAs meet the *Minimum Elements of HIA* and exemplify HIA best practices.

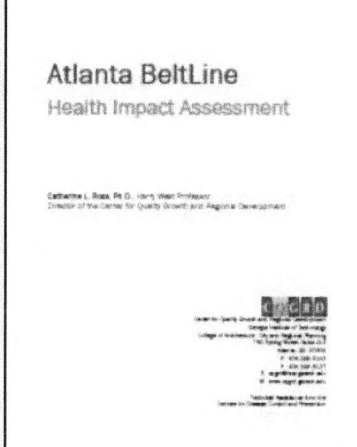

Atlanta Beltline Health Impact Assessment
<u>Strengths</u>
- adherence to HIA standards and methodology
- included description of screening process in the HIA
- provided logic model framework
- assessed comprehensive health endpoints
- sources of evidence and methodology sound and of high quality
- included survey and results in the HIA
- tabular summary of HIA showing key findings, affected populations, recommendations, and the categories of health impacts attributed to each recommendation
- identified lessons learned

Healthy Corridor for All: A Community Health Impact Assessment of Transit-oriented Development Policy in St. Paul Minnesota
<u>Strengths</u>
- adherence to HIA standards and methodology
- identified core values that guided the HIA
- included a description of methodology used at each step of the HIA, including screening
- provided a logic model framework
- used a Rules of Engagement Memo
- sources of evidence and methodology sound and of high quality
- provided a detailed description of data sources, including geographic scale of the data, and methodology used in data analysis

Figure 38. Model HIAs from the Land Use sector.

Waste Management/Site Revitalization

The four HIAs in the Waste Management/Site Revitalization sector were conducted to assess the impacts of a number of waste-related projects and policies (Table 10). As described previously, the effectiveness of these HIAs in influencing the decisions at hand was evaluated by the HIA reviewers using four measures of effectiveness defined by Wismar et al. (2007) and information obtained via an internet search. As such, the measures of effectiveness noted in Table 10 are subjective and may not reflect the true effect of the HIA on the decision-making process. Likewise, the measures of effectiveness noted in the table, do not necessarily reflect the overall effectiveness of the HIA, but rather the HIA's effect on the decision. For a more detailed discussion of the four measures of effectiveness utilized here and measures of overall HIA effectiveness, see *HIA Effectiveness*.

Table 10. Waste Management/Site Revitalization Decisions Informed by Reviewed HIAs

ID	Title	Decision-making Level	Decision Type	HIA Scope/Summary	HIA Rigor	Effect of HIA on Decision-making
4	Health Impact Assessment of NRMT's Request for a Special Use Permit	County	Project	Address the health impacts of the proposed dirty materials recovery facility.	Intermediate	Direct effectiveness
25	Concord Naval Weapons Station Reuse Project Health Impact Assessment	Local	Project	Analyze how the alternatives being considered for the CNWS Reuse Project would help realize health and well being benefits or potentially lead to negative health outcomes.	Intermediate	Direct effectiveness
33	Neenah-Menasha Sewerage Commission Biosolids Storage Facility, Greenville, WI	County	Project	Review potential health concerns and propose methods to reduce those risks with the building of a biosolids storage facility.	Rapid	Undetermined
74	Assessment of Open Burning Enforcement in La Crosse County	County	Program/Policy	Determine the potential health impacts of creating a uniform open air burning policy within La Crosse County.	Rapid	Undetermined

Appendix G, Table G-4 provides a more detailed look at each of these HIAs, including sources of evidence, impacts/endpoints, pathways of impact, characterization of impacts, decision-making outcome, evidence of HIA effectiveness, and evaluation of whether the *Minimum Elements of HIA* were met. Figure 39 provides a dashboard of summary statistics for these HIAs and Figure 40 highlights one of the tools utilized in analysis.

76

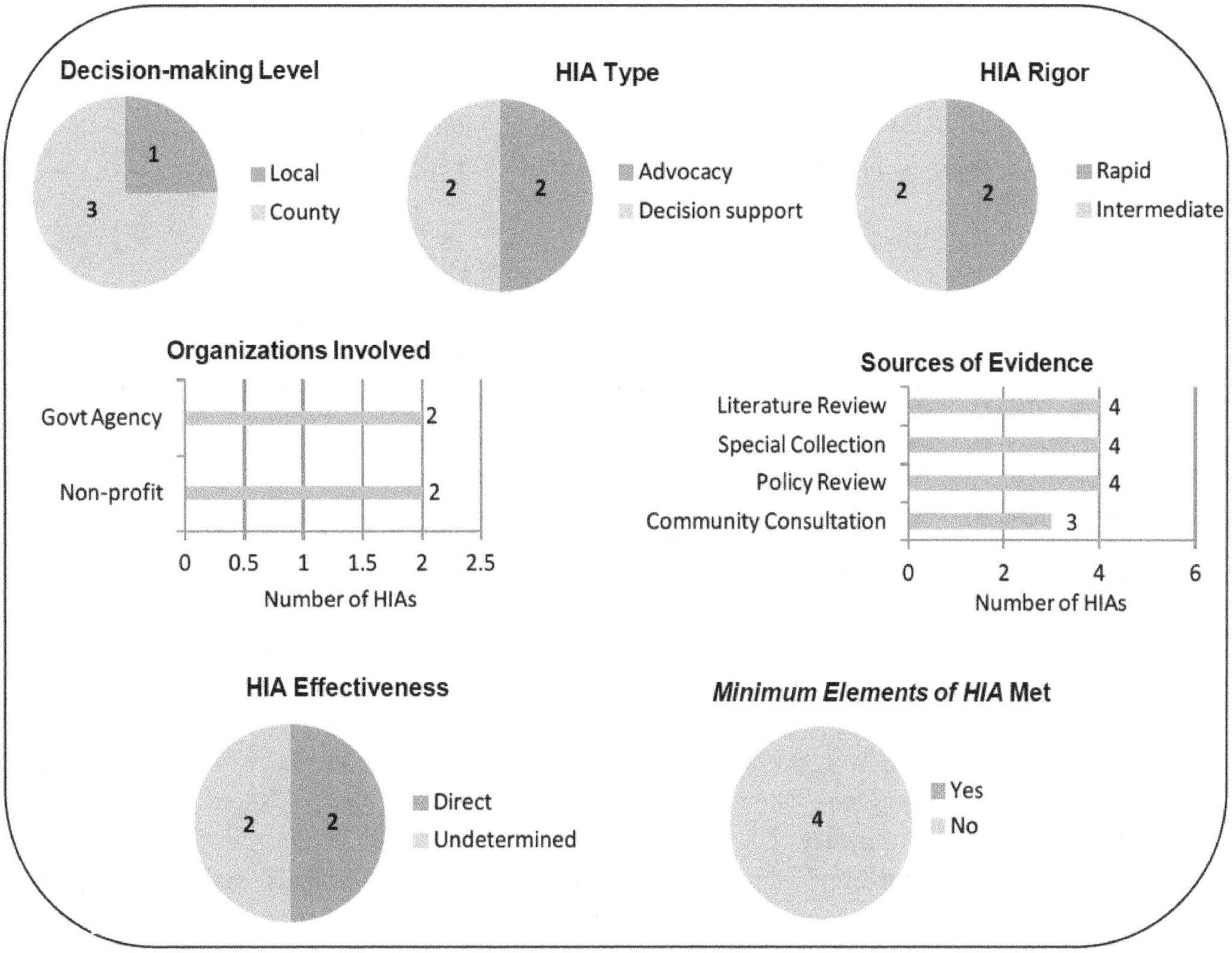

Decision-making Level

1
3

- Local
- County

HIA Type

2   2

- Advocacy
- Decision support

HIA Rigor

2   2

- Rapid
- Intermediate

Organizations Involved

Govt Agency — 2
Non-profit — 2

0   0.5   1   1.5   2   2.5
Number of HIAs

Sources of Evidence

Literature Review — 4
Special Collection — 4
Policy Review — 4
Community Consultation — 3

0   2   4   6
Number of HIAs

HIA Effectiveness

2   2

- Direct
- Undetermined

Minimum Elements of HIA Met

4

- Yes
- No

Figure 39. Dashboard of summary statistics for reviewed HIAs in the Waste Management/Site Revitalization sector.

In the
Spotlight

Living Wage Calculator

The Living Wage Calculator was used in the Concord Naval Weapons Station Reuse Project HIA to calculate the living wage for the city of Concord and compare this with the wages of current jobs in the area, as well as the wages of jobs likely to be created by the project. Using this tool, the HIA was able to show that over 50% of the created jobs would pay workers less than the living wage and that the amount of affordable housing proposed in the project alternatives would not meet the demand created by these wages. This led to recommendations to 1) adopt a Living Wage Ordinance that ensures that new jobs pay residents enough for them to live in Concord and have a reasonable quality of life, and 2) match the cost of new housing to the projected wages of new jobs to meet the affordable housing need. The Living Wage Calculator is available at: http://livingwage.mit.edu/

Figure 40. Tool spotlight: Living Wage Calculator

All 4 HIAs in the Waste Management/Site Revitalization sector examined environmental impacts or endpoints, 3 examined economic endpoints, and 3 utilized GIS.

None of the HIAs in the Waste Management/Site Revitalization sector met the *Minimum Elements of HIA*, so no model HIAs are provided for this sector.

Current State-of-Science in HIA Community of Practice

Based on the HIAs reviewed, conclusions can be drawn about the current state-of-science in the HIA community of practice, including the implementation of HIAs, best practices, and areas for improvement.

HIA Implementation

The HIA Review revealed several trends in implementation of HIAs within the four chosen sectors that can be safely extrapolated to the greater community of practice.

Use of HIA to Inform Decision-making

Based on the HIAs reviewed, it is evident that HIAs are being used with increased frequency to bring health to the decision-making process. This trend is consistent with trends seen in the overall community of practice (i.e., the use of HIAs in the U.S. is on the rise; Health Impact Project 2013). While the reviewed HIAs were implemented consistently across the four sectors to inform decisions at the local level, they were used less frequently to inform decisions at the county, state, and federal level.

Implementation of the HIA Process

Implementation of the six-step HIA process varied greatly among the reviewed HIAs, leading to large disparities in rigor and quality.

- *Screening* – Documentation of the screening process was lacking throughout the HIAs reviewed, making it difficult to discern what factors went into making the decision to conduct the HIA. This is consistent with trends seen in the overall community of practice (National Research Council 2011). But as the National Research Council (2011) notes, while the reasons for proceeding with an HIA are often unclear, still less is known about the reasons that lead to decisions not to proceed with an HIA.

- *Scoping* – Documentation of the scoping process was not consistent among HIAs and often lacked details of the overall HIA plan (including research questions to be answered and rationale for issues selected for inclusion in the HIA), despite readily-available guidance and even scoping templates developed by Human Impact Partners (http://www.humanimpact.org/component/jdownloads/finish/13/5) and others.

Organizations involved in conducting HIAs vary, although state and local government health agencies were most commonly involved in carrying out the assessments included in the HIA Review. While HIAs are typically successful in bringing health to the table in decisions outside of traditional health-related fields, the HIA field of practice could be advanced with successful implementation of HIAs by non health-related organizations.

The rigor of the HIAs vary drastically due to the scale and complexity of the decisions being assessed, the timing of the HIA relative to the decision-making process, and disparities in the breadth of the HIA scope (i.e., the number of impacts to assess), depth of impact assessment, extent of data collection and analysis (including input from stakeholders), time and resources allocated to complete the HIA, and likely the skills and expertise of the HIA team, although the latter was not assessed in the HIA Review.

It is often infeasible to examine all the impacts of a decision, making it necessary to prioritize the impacts to be considered in the HIA. Prioritization can be based on a number of factors, but those used most frequently in the reviewed HIAs included stakeholder/community input, literature and research, impact on health and relevance to project/decision interests, and equity of impacts. While considerable evidence exists regarding the various pathways through which health can be affected and the value added of logic frameworks and pathway diagrams in identifying links between proposed decisions and health, these tools were inconsistently applied in the scoping process of the HIAs.

- *Assessment* – The depth and defensibility of the evidence used in HIA is crucial to the effectiveness of impact assessment. Existing data, tools, and models used in the assessment step of the reviewed HIAs were gathered from a variety of resources. The U.S. Census Bureau was the primary resource used in the HIAs for demographics and background data (e.g., social, economic, housing, and educational attainment data), while health data were most commonly gathered from state, county, or local health departments or one of the various health surveys conducted by the CDC.

In addition to the use of existing data, tools, and models, collection of primary data is also a critical component of HIA. A variety of special collection methods were used in the HIAs to acquire new data, often at the "local" level (i.e., the level of the project or decision). Of the special collection methods utilized, those that involved and/or solicited information from stakeholders or the community and GIS or other mapping techniques were used most often.

Several deficiencies were found in the HIAs related to evidence defensibility. These include: lack of clear/cited supporting evidence, marginal or moderate quality of evidence, no identification of assumptions and/or limitations of the assessment, and lack of transparency in the synthesis of evidence.

A critical component of the assessment process is developing a profile of baseline conditions that includes data on health outcomes and determinants of health. These baseline conditions are necessary in order to identify disparities in existing conditions and predict future conditions if the decision is implemented, yet the extent of the profiles created in some of the HIAs was very limited and in others, it was missing completely (n=18).

Overall, the potential health impacts from decisions in the 81 reviewed HIAs were extensive, but on the individual HIA level, the depth of health impacts considered in assessment varied greatly. In addition to identifying potential health impacts, the assessment step in the HIA process also includes characterizing the direction, magnitude, likelihood, distribution (i.e., equity), and permanence of impacts via qualitative and quantitative analysis. However, the HIA Review found that impact characterization rarely considered likelihood, magnitude, or permanence; characterization of impacts primarily involved considerations of direction and distribution/equity.

In addition, most HIAs qualitatively characterized impacts; the use of quantitative analysis was lacking. The participatory research aspects of HIA (i.e., stakeholder and community involvement) lend themselves to qualitative analysis, and in HIA, unlike other forms of risk assessment, impacts that are able to be characterized qualitatively can often be more relevant to the decision at hand than those impacts that are able to be quantified (National Research Council 2011). While qualitative characterization is acceptable, and many times warranted due to a lack of available scientific research, local data, time, and resources, there is also value added in the use of quantitative estimates when the process allows. When empirical research exists linking a health determinant to a specific health outcome and the data, time, and resources are available, quantification of health impacts via modeling, forecasting, and other tools (e.g., exposure-response relationships) can provide depth and defensibility to the impact assessment, allowing estimates of magnitude of impact and in some cases, permanence (or severity) of impact when compared against established threshold values (National Research Council 2011).

- *Recommendations* – Recommendations provided in HIA typically include alternatives or modifications to the decision to promote positive health impacts or minimize negative health impacts, and/or direct mitigations for negative health impacts; in some cases, support for or opposition to the decision being assessed is also offered. In most cases, HIAs also include some sort of prioritization process to identify impacts for which recommendations will be offered and/or identify which of the developed recommendations to offer or prioritize for action.

In the recommendations stage of the HIA process, it is suggested that an implementation plan be prepared for the developed recommendations that includes information such as parties responsible for implementation, timeline, and links to indicators to be monitored. However, implementation plans or strategies for recommendations were found in only 10% of the HIAs reviewed.

- *Reporting* – Reporting and communicating the results of HIA are crucial to informing the decision being evaluated, yet only 4 of the 81 reviewed HIAs made mention of or included in the HIA report a communication plan or strategy for reporting and disseminating the findings and results to the appropriate audiences, although communication plans could have been developed and documented separately from the HIA Report.

Preparation of a publicly-accessible HIA report documenting the HIA and its results is one component of the reporting phase. When identifying HIAs for inclusion in the HIA Review, there were multiple instances of HIA reports not being readily accessible, which precluded those HIAs from the review. The reporting phase of the HIA process not only calls for the HIA report to be publicly accessible, but also transparent. Of the 81 HIAs reviewed, over 35% lacked transparent documentation of the processes, methods, findings, sponsors, funding source(s), and/or participants and their respective roles. There were a number of HIAs, however, that went above and beyond to ensure that the documentation was transparent, including detailed documentation of assessment methodologies, techniques, and models; criteria for data aggregation; geographic units of analysis (i.e., geographic area and scale); confidence estimates; and supporting documentation in the HIA report.

In addition to preparing the HIA report, reporting also involves communicating and disseminating the findings and recommendations of the HIA to inform stakeholders and decision-makers. Methods for communicating the HIA results to decision-makers and stakeholders took many forms in the reviewed HIAs, including dissemination of the HIA Report, factsheets, presentations, press releases, public and/or stakeholder meetings, public testimony, lobbying, personal communication, and publication on websites and peer-reviewed journals.

- *Monitoring and Evaluation* – This step of the HIA process was almost completely lacking in the 81 HIAs reviewed. Unfortunately, this is not a trend unique to this subset of HIAs (Wismar et al. 2007; National Research Council 2011). Of the three forms of evaluation called for in the monitoring and evaluation phase of HIA (i.e., process evaluation, impact evaluation, and outcome evaluation), process evaluation was found in only 5 HIAs (although process evaluation could have been performed and reported separately) and proposed plans for impact and/or outcome evaluation were present in only 29 of the HIAs.

Because both impact evaluation (i.e., monitoring the effect of the HIA on the decision-making process and final decision) and outcome evaluation (i.e., monitoring the effect of decision implementation on health) are carried out after completion of the HIA, these measures are not documented as part of the HIA Report. During internet searches conducted throughout the duration of the HIA Review, little to no documentation of these evaluations were found. Monitoring and evaluation is a definite area for improvement in the HIA community of practice.

Stakeholder/Community Engagement

The engagement of stakeholders and the community in the HIA process varied greatly in the HIAs. While stakeholder and community engagement in each step of the HIA process is ideal, this was rarely witnessed in the HIA Review. In fact, of the 81 HIAs, almost 20% (n=15) did not engage stakeholders or the community at all in the HIA process. Of those HIAs with no stakeholder and/or community involvement, ten were desk-based HIAs, which by definition do not include stakeholder or community involvement, yet stakeholder engagement and community empowerment are objectives of HIA (North American HIA Practice Standards Working Group 2010; National Research Council 2011; Human Impact Partners 2012). Engagement of the community, and in particular vulnerable populations, is critical to ensuring equity is promoted in HIA and can help illuminate issues and existing conditions that might not be readily apparent to those outside the community. Engagement of decision-makers as stakeholders in the process is also beneficial and ensures that the recommendations offered are realistic, practical, and able to be implemented within the purview and authority of the decision-makers.

Among the HIAs with a stakeholder or community involvement component, the level and quality of stakeholder participation varied greatly. In many of these HIAs, stakeholder input was solicited to inform the scoping and assessment steps of the process (e.g., identify issues of interest and areas of concern for the community and stakeholders, identify populations and vulnerable groups that might be affected by the decision, gather local knowledge regarding community health and existing conditions, etc.), but the stakeholders themselves were not involved in the actual HIA decision-making. However, there were a number of HIAs (n=15) that did engage stakeholders in the decision-making process, usually via a role on an advisory or steering committee, and in a handful of HIAs (n=4), stakeholders actually oversaw or guided the HIA process and were engaged as decision-makers in equal partnership with the HIA team or as the primary decision-makers in the process.

Due to the variety of decision contexts, capacities, and stakeholder groups that could potentially be involved in HIA, there is no single approach to engaging stakeholders in the process, however, HIA guidance does call for HIA practitioners to employ "deliberative methods" of stakeholder engagement (North American HIA Practice Standards Working Group 2010; National Research Council 2011). Deliberative stakeholder engagement "makes a difference, is transparent, has integrity, is tailored to the circumstances, involves the right number and types of people, treats participants with respect, gives priority to participants' discussions, is reviewed and evaluated to improve practice, and keeps participants fully informed" (Warburton et al.

2008). Increased rigor in the engagement of stakeholders and the community in the HIA process is an area for improvement that could help advance the HIA community of practice.

HIA and Environmental Impacts

Characterization of environmental or ecosystem impacts was present in 50 of the 81 HIAs reviewed, but with the exception of the 5 HIAs conducted in support of environmental impact assessments, these HIAs primarily examined impacts via the air quality pathway. Increased consideration of environmental impacts via other pathways could help advance the effectiveness of HIA in predicting health impacts. For example, the consideration of impacts on soils, water quality, water quantity (e.g., stormwater runoff, flooding, groundwater and drinking water recharge, etc.), vegetation and green space, habitat (e.g., habitat quality, loss, fragmentation), wildlife (e.g., aquatic and terrestrial species), sustainability and stewardship, and temperature and climate change (e.g., the urban heat island effect, greenhouse gases) could, in some instances, contribute to a more robust impact assessment. The EPA's Eco-Health Relationship Browser (http://www.epa.gov/research/healthscience/browser/introduction.html) illustrates some of the scientific evidence linking human health to some of these ecosystem services (i.e., benefits supplied by nature).

In addition to broader consideration of environmental and ecosystem impacts in HIAs, HIA can also be used as a tool for incorporating health considerations into environmental impact assessments. As mentioned previously, NEPA requires the U.S. government to consider environmental and human health effects prior to undertaking any major federal action that will significantly affect the quality of the human environment, but historically the examination of health effects in EIA is rare and narrowly focused (Bhatia and Wernham 2008). The EIA and HIA processes are similar in many regards. Like HIA, EIA examines potential effects of the decision under consideration, provides opportunity for stakeholder involvement, and develops recommendations to minimize risks and maximize benefits. If health effects were considered in EIA to the same extent that environmental effects are considered, the resulting health effects analysis would be consistent with conducting an HIA (National Research Council 2011). Therefore, HIA can potentially provide a means of complying with the NEPA requirement to include human health considerations (i.e., direct, indirect, and cumulative human health effects) in analysis, and in fact, HIAs have been successfully incorporated into the EIA process in the U.S. on multiple occasions (Bhatia and Wernham 2008; National Research Council 2011). When health effects have not been adequately considered in the EIA process, HIA has provided a means of bringing those issues to light and informing the EIA. In other cases, rather than performing the HIA separately from the EIA, it has actually been integrated into the EIA process. In these integrated EIA/HIAs, health expertise is included on the team conducting the EIA and health is integrated throughout the EIA process (e.g., stakeholder involvement, analysis of impacts, development of recommendations, and reporting). Whether it results in a separate standalone report or is integrated into the EIS, HIA can improve the consideration of health in EIA.

Adherence to *Minimum Elements and Practice Standards*

Less than 20% (n=13) of the HIAs examined met all the *Minimum Elements of HIA*. Elements most often missing in the HIAs included complete characterization of impacts (direction, magnitude, likelihood, distribution, and permanence), inclusion of a plan for monitoring, and transparency in documentation. Adherence to these *Minimum Elements and Practice Standards* is crucial for advancing the use of HIAs in the U.S and is a definite area for improvement in the HIA community of practice overall.

Effectiveness of HIA

As part of the HIA Review, evaluations of HIA effectiveness were conducted by the reviewers using information able to be obtained via an internet search. As such, it should be noted that these evaluations of effectiveness are subjective and may not reflect the true effectiveness of the reviewed HIAs. In many cases it was difficult to discern from the internet, the effectiveness an HIA had in influencing the decision at hand. Effectiveness could not be determined for almost 40% (n=31) of the reviewed HIAs, but for those HIAs for which measures of effectiveness were made, 60% showed direct effectiveness (i.e., the decision was dropped, modified, or postponed as a result of the HIA), 32% showed general effectiveness (i.e., the HIA was considered, but did not result in modification to the decision), 6% showed no effectiveness, and 2% showed opportunistic effectiveness (i.e., the decision was going to be carried out regardless of the HIA).

While the high degree of direct effectiveness seems to support the use of HIA in decision-making, one could hypothesize that effectiveness could not be determined for the 31 HIAs because there was no direct effect on the decisions in those cases (i.e., documentation of the decision-making process available on-line did not attribute any portion of the final decision to the HIA). Beyond influencing the decision-making outcome, a number of other measures of effectiveness were observed in the reviewed HIAs. These included raised awareness of health and related issues; the introduction of health into discussions where health was typically absent (i.e., informing decision-making); engagement of community members and stakeholders in decisions that affect them; interdepartmental, interagency, and even intersector collaborations; and relationship and capacity building within the community. If HIA practitioners are to make the case for the use of HIA in decision-making, the value and effectiveness of this tool in informing decisions and protecting and promoting human health needs to be established.

Best Practices

Best practices were identified within various HIAs in the HIA Review and are compiled here for future use, in an attempt to advance the HIA field of practice and reduce disparities in the quality and rigor of HIA. A list of these best practices is provided in Table 11, and where appropriate, the best practices are supported by examples from the reviewed HIAs in the figures that follow.

Table 11. Best Practices in HIA

Best Practice	Description
Adherence to *Minimum Elements and Practice Standards for Health Impact Assessment* (North American HIA Practice Standards Working Group 2010) or similar criteria developed by the National Research Council (2011)	Adherence to the *Minimum Elements* ensures the elements that distinguish HIA from other assessment processes are met (i.e., screening; scoping; stakeholder engagement; establishment of baseline conditions; judgement of impact direction, magnitude, likelihood, distribution, and permanence using best available evidence; transparent and context-specific synthesis of evidence, including assumptions, strengths, limitations, and uncertainties; identification of recommendations that promote health; monitoring plan proposal; and transparent, publicly-accessible documentation). Adherence to the *Practice Standards* allows benchmarks for effective HIA practice, rigor, and quality to be met. While the National Research Council (2011) criteria and *Practice Standards* are intended to reflect an ideal of practice rather than rigid requirements for HIA implementation, the need and rationale for deviations from these should be clearly articulated.
HIA as a Tool for Environmental Impact Assessment	Use of HIA in the environmental impact assessment process can provide a means of meeting the NEPA requirement to include human health effect considerations. Whether it results in a separate standalone report or is integrated into the EIS, HIA can improve the consideration of health in EIA. Note: If the HIA was completed separately from the Environmental Impact Statement (EIS), the HIA Report should be included as an appendix to the EIS, with appropriate health information incorporated into the body of the EIS, as needed for transparency in documentation.
Equity Promotion	Promotion of equitable health outcomes and empowerment of vulnerable communities is central to HIA. In addition to judging the distribution and equity of predicted health impacts in the HIA process, equity can be promoted in a number of other ways (Heller et al. 2013): • employing HIA to inform decisions that are identified by, or relevant to, vulnerable populations; • promoting community ownership and participation in the HIA; • engaging vulnerable populations in the decision-making process; • identifying recommendations that result in equitable health outcomes; • communicating findings and recommendations of the HIA to the community; and • monitoring the impacts of implemented decisions on community health.

Table 11. Continued

Best Practice	Description
Documentation of Screening and Scoping	Clear documentation of the selection process for the HIA, including screening criteria/factors that went into making the decision to perform the HIA; and HIA scope, including participants and their roles, the issues prioritized for inclusion (and the rationale), research questions to be answered, methods to be employed, and the timeline for completion. The National Research Council (2011) provides an excellent summary of the recommended outputs for both of these steps in the HIA process. See Figure 41 for an excerpt of the Scoping Worksheet from the Daniel Morgan Road Diet and Restriping HIA.
Rules of Engagement Memo/ Memorandum of Understanding	A Rules of Engagement Memo or Memorandum of Understanding establishes groundrules for HIA participants, and outlines expected outcomes, participant responsibilities, and protocols for information sharing throughout the project. See *Appendix H* for the Rules of Engagement Memo from the St. Paul Healthy Corridors for All HIA.
Communication/Reporting Plan	Communicating the findings and recommendations of the HIA is crucial to informing the decision being evaluated. Identification and documentation of communication and reporting strategies (e.g., types of communication, methodology, audience, and timing) early on, helps to ensure effective communication throughout the HIA process. See Figure 42 for the communication methods used in the HIA on Wind Energy Development.
Stakeholder Involvement	Stakeholders, including decision-makers and the community, should be engaged at every step of the HIA process to inform and provide input into the findings and results of the HIA. Stakeholders may also play a more substantial role in the HIA process and actually conduct the HIA, such as in the case of community-led HIAs. No matter the level of stakeholder involvement, HIA practitioners should employ "deliberative methods" to engage stakeholders in the process. The principles of deliberative stakeholder engagement are that "the process: · makes a difference; · is transparent; · has integrity; · is tailored to the circumstances; · involves the right number and types of people; · treats participants with respect; · gives priority to participants' discussions; · is reviewed and evaluated to improve practice; and · that participants are kept informed." (Warburton et al. 2008) See *Appendix I* for a summary of stakeholder involvement opportunities in each step of the HIA process, and the St. Paul Healthy Corridors for All HIA for an example of deliberative stakeholder engagement.

Table 11. Continued

Best Practice	Description
Transparent Literature Search/ Review Documentation	Clear, thorough description of the literature review process, including search terms and bibliometric databases searched. Documentation may also include identification of the number of articles yielded by the literature search and the final number of articles used as evidence base. See Figure 43 for a tabular summary of literature review results from the HIA of the Transform Baltimore Zoning Code Rewrite. In addition to this summary, a quality review of the literature was also conducted (see Quality of Evidence Evaluation best practice) and the terms used in the literature search documented in the HIA Report.
Use of Best Available Data (Qualitative & Quantitative)	The use of best available data – both qualitative and quantitative – is necessary to provide a strong evidence base on which to predict potential health impacts. When empirical research exists linking a health determinant to a specific health outcome and the data and time are available, quantification of health impacts via modeling, forecasting, and other tools (e.g., exposure-response relationships) can provide depth and defensibility to the impact assessment, allowing estimates of magnitude and, in some cases, permanence (or severity) of impacts. Regardless of whether qualitative or quantitative analysis is utilized, the methodology and analytical approach should be clear and rigorous.
Quality of Evidence Evaluation	An evaluation of the quality or strength of evidence on which the impact assessment is made provides transparency and defensibility in HIA. Documentation of that evaluation includes identification of criteria used in making the determinations of quality/strength and reporting the results of the evaluation. The quality review conducted in the HIA of the Transform Baltimore Zoning Code Rewrite, for example, was based on whether the papers used an appropriate study design, whether they adequately controlled for confounding by socioeconomic status, and whether they used self-reported as opposed to externally measured variables; studies were then categorized into one of three quality groups – good, fair, or poor –based on these criteria (see Figure 43). Other HIAs have used ratings based on the number and strength of studies used as evidence (e.g., < 5 studies and claim consistent with public health principles, 5 or more studies of weak or moderate quality, 5-10 strong studies and/or data analysis, 10+ strong studies). See Figure 44 for the hierarchy of evidence used in the HIA on Wind Energy Development.
Identification of Data Gaps	Clear identification of data gaps in the HIA report, especially as a stand-alone presentation (e.g., section or table) provides transparency in reporting. See Figure 45 for data gaps identified by health effect category in the HIA for the Proposed Coal Mine at Wishbone Hill.
Use of Existing Tools, Methods, Standards, and Metrics	There is no need to start from scratch. Use of existing tools, methods, standards, and metrics provides efficiency and consistency in HIA.

Table 11. Continued

Best Practice	Description
Adaptation of Existing Tools and Methods	While use of existing tools and methods is ideal, there may be the need to tailor these items for the problem at hand. For instance, Humboldt County modified the existing Healthy Development Measurement Tool (HDMT), which was created for an urban environment, to create a version of the tool that allows health to be considered in development decisions in rural environments.
Detailed Documentation of Data and Methodology	Detailed documentation of the data (e.g., key input data, data variables, data sources, geographic scale, etc.) and methodologies used in analysis provide transparency and defensibility in the HIA. See Figure 46 for a description of key input data from the Health Effects of Road Pricing in San Francisco HIA; See Figure 47 for a detailed description of data analysis methodology from the St. Paul Healthy Corridors for All HIA.
Geographic Information Systems (GIS)	GIS allows geographically-referenced data to be displayed in visually-pleasing maps, but can also be used to analyze and interpret geographically-referenced data to reveal relationships, patterns, and trends. GIS can be used to identify spatial disparities in health outcomes, evaluate health determinants and outcomes in a geographic context (e.g., evaluate proximity measures identified for health determinants), identify environmental justice (EJ) communities, link health and environmental data in geographical modeling and analysis, and even combine incompatible spatial data (Gotway and Young 2002; Young et al. 2009; AAG 2012; Deganian and Thompson 2012).
Impact Pathways/Logic Frameworks	The use of impact pathways and logic frameworks to identify links between the proposed decision and health is value added. Of additional value is clear identification of pathways of exposure to contaminants and pollutants. See Figure 19 for a pathway diagram from the HOPE VI to HOPE SF HIA; See Figure 20 for a logic framework from the Health Effects of Road Pricing in San Francisco HIA showing pathways between road pricing and health; See Figure 48 for identified exposure pathways from the Beaufort Sea and Chukchi Sea Oil and Gas Lease HIA.
Clear Summary of Impact Assessment	While characterization of impacts is essential in HIA, the use of tables and graphics to summarize the findings of impact assessment provides added transparency. Summaries of impact direction, magnitude, likelihood, permanence, and distribution can be provided, as well as summaries of differential impacts between alternatives being analyzed. See Figure 49 for a summary of differential impacts of alternatives in the Trenton Farmers Market HIA; See Figure 50 for a summary of health impact assessment findings from the Daniel Morgan Road Diet and Restriping HIA.

Table 11. Continued

Best Practice	Description
Confidence Estimates/Assessments of Uncertainty	Confidence estimates assess uncertainties in the assumptions, parameters, and methodologies on which the health impact characterization is based. Uncertainties can include limitations, gaps, or weaknesses in individual evidence sources used in characterization, considerations of whether measures used in assessment are reliable proxies for the intended factor, whether an existing model or tool from the literature can be generalized to the study area or population of interest, etc. Confidence estimates can entail qualitatively identifying the uncertainty, explaining how the data inputs used in analysis may vary from actual (be over- or under-estimated), and describing the influence of that variation on the health impact conclusions, or quantitatively assessing uncertainty using methods such as sensitivity analysis (Bhatia 2011, National Research Council 2011). Characterization and management of uncertainty in assessment provides transparency and defensibility. See Figure 51 for uncertainty factors and confidence estimates from the Health Effects of Road Pricing in San Francisco HIA.
Prioritization Process for Recommendation Development/Action	A prioritization or ranking process can be used to identify which impacts require recommendations and/or which of the developed recommendations to offer for action. Prioritization can be based on a number of factors, but prioritization methods commonly used include stakeholder/community input, literature and research, impact on health and relevance to project/decision interests, and equity of impacts. See Figure 52 for a ranking process used in the HIA for the Proposed Coal Mine at Wishbone Hill and *Appendix J* for a unique risk assessment technique used to prioritize impacts for management actions in the Point Thomson Project HIA.
Recommendations That Meet Established Feasibility Criteria	Human Impact Partners (2012) provides the following criteria for recommendations: • Responsive to predicted impacts • Politically feasible • Specific and actionable • Economically efficient • Experience-based and effective • Do not introduce additional • Enforceable negative consequences • Able to be monitored and enforced • Relative to the authority of • Technically feasible decision-makers
Implementation Plan for Recommendations	Recommendations identified in HIAs are only effective if they are implemented. Development of an implementation plan for identified recommendations identifies information such as parties responsible for implementation, audience for the recommendation, timeline for implementation, and links to indicators that can be monitored. See Figure 53 for a summary of recommendation strategies and decision-makers from the Page Avenue HIA; See Figure 54 for a summary of recommendations and indicators from the Highway 99 Sub-Area Plan HIA.

Table 11. Continued

Best Practice	Description
Clear/Transparent HIA Report	Clear, transparent documentation of the HIA process should include documentation of sponsors; funding sources; timelines; participants and their roles; what tasks and activities were undertaken at each step of the process; data and methodologies; results of analysis; conclusions and findings; recommendations; and assumptions and limitations. Tables and figures should be used in the body of the report to illustrate key information and appendices used to detail supporting documentation. See Figure 55 for an example of transparent documentation of one stage in the Eastern Neighborhoods Community HIA.
Process Evaluation	Process evaluation involves an evaluation of HIA quality and defensibility and should examine how the HIA process was carried out, including who was involved, strengths and weaknesses of the HIA, successes and challenges, effectiveness in meeting HIA objectives and established practice standards, engagement and communication with stakeholders, and lessons learned. During the scoping phase, steps should be taken to consider how process evaluation can be built into the HIA process. Example evaluation questions are available from Human Impact Partner (2012).
Monitoring Plan – Impact and Outcome Evaluation	Impact and outcome evaluation are both carried out after completion of the HIA; however, the HIA should include or, at a minimum, acknowledge plans for monitoring the impacts of the HIA on the decision and decision-making process (i.e., impact evaluation) and the impacts of the decision implementation on health determinants and outcomes (i.e., outcome evaluation). If impact and/or outcome evaluation is infeasible, the HIA should discuss the limitations that are preventing monitoring from occurring (e.g., the length of time between implementation of the decision and changes in health outcomes, the presence of multiple contributing factors to health outcomes, etc.). Per Human Impact Partners (2012), the essential elements of a monitoring plan include: ◦ Goals ◦ Resources to conduct and report monitoring ◦ Identification of the outcomes, processes, impacts, and indicators to be monitored ◦ Process for collecting meaningful and relevant information (e.g., baseline and long-term data) ◦ Defined roles for individuals or organizations involved in monitoring ◦ Criteria or triggers for action, if agreed upon mitigations or recommendations are not met ◦ Process for reporting monitoring (methods and results) and making them publicly available ◦ Process for learning, adapting, and responding to monitoring results See Figure 56 for a monitoring plan from the Rental Assistance Demonstration Project HIA.

APPENDIX B: SCOPING WORKSHEET

Project: Road Diet/Re-Striping of Daniel Morgan Avenue (DMA), Spartanburg, SC
Health Determinant: Traffic Safety
Geographic Scope: Stretch of DMA included in proposed project

Existing Conditions	Impact Research Questions	Indicators	Data Sources	Methods
What is the existing traffic count on DMA?	How will the new pedestrian and bicycle lanes affect the number of cars on DMA?	Average Annual Daily Traffic count	SCDOT/SPATS	Pneumatic road tube
How many pedestrians and bicyclists travel on DMA each day?	What is the anticipated change in the number of bicyclists and pedestrians?	Pedestrian and bicycle counts	PAL	Manual observation
What is the current motor vehicle collision rate on DMA?	How will the proposed road re-configuration affect motor vehicle collision rates?	Number and locations of collisions	SCDOT Public Safety Crash Data	Review of incident reports
What is the current Bicycle Level of Service (BLOS) on DMA?	How will the Bicycle Level of Service (BLOS) be affected by the proposed bicycle lane?	Bicycle Level of Service (BLOS)	SPATS	Map review
What is the average rate of speed on DMA?	To what extent can it be anticipated that installing the road diet will reduce vehicle speed?	Speed	SCDOT	Radar

Project: Road Diet/Re-Striping of Daniel Morgan Avenue (DMA), Spartanburg, SC
Health Determinant: Physical Activity
Geographic Scope: Stretch of DMA included in proposed road diet Spartanburg/Half-mile radius around DMA/ City of Spartanburg/Spartanburg zip codes 29303 and 29306 (depending on the data)

Existing Conditions Research Questions	Impact Research Questions	Indicators	Data Sources	Methods
How many pedestrians and bicyclists travel on DMA each day?	How will the new pedestrian and bicycle lanes affect the number of pedestrians and bicycles on DMA each day?	Pedestrian and bicycle count	PAL	Manual observation
How many bicycle riders are there each day within a half-mile radius around DMA?	What is the expected increase in the number of bicycle riders within a half mile radius of DM Avenue each day?	Number of bicycles on bicycle racks within a half-mile radius of DMA	PAL	Manual observation
What is the utilization of PAL's bike lending program?	What is the anticipated impact on usage of the PAL bicycle lending program?	Number and trend of users of bike lending program	Documentation through PAL bike lending records	Record review
How many minutes/day of physical activity do residents of the zip codes surrounding DMA get on average?	How will the average number of minutes/day of physical activity be affected?	Minutes/day on average of physical activity	Spartanburg Bicycle Pedestrian Master Plan Survey, 2009	Secondary Data Analysis by zip code

Figure 41. Best Practice: Documentation of Screening and Scoping – a scoping worksheet (Source: Daniel Morgan Road Diet and Restriping HIA).

	Audience	Communications
Press	General Public	Announcement of meetings, report release
Website	General Public	Description of process, announcement of meetings, updates, invitations to provide feedback, links to electronic versions of report
Public meetings	General Public in North Central/ Eastern Oregon	Opportunity for community members to share thoughts, questions, concerns
General Listserv	Interested parties	Announcement of meetings, survey, progress updates, report release
Steering Committee Listserv	Steering committee	Meeting logistics, announcement of meetings, training opportunities, information sharing
Personal Communications	Interested parties	Response to individual calls, emails, and letters

Figure 42. Best Practice: Communication/Reporting Plan – methods of communication (Source: HIA on Wind Energy Development in Oregon).

TABLE 5-1: LITERATURE REVIEW FINDINGS FOR BUILT ENVIRONMENT FEATURES AND OUTCOMES OF INTEREST

CITATION	STUDY QUALITY	BUILT ENVIRONMENT INDICATORS					FOOD ENVIRONMENT INDICATORS							PRIMARY HEALTH OUTCOMES OF INTEREST					
		ALCOHOL SALES OUTLETS	CPTED	MIXED USE	PEDESTRIAN SCALE	TOD	SUPERMARKETS	GROCERY STORES	COMMUNITY GARDENS/URBAN AGRICULTURE	CORNER & CONVENIENCE STORES	FULL SERVICE RESTAURANT	FAST FOOD	BARS & TAVERNS	CRIME	PEDESTRIAN INJURY & FATAL CRASHES	PHYSICAL ACTIVITY	HEALTHY DIET & NUTRITION	OBESITY & BMI	OBESITY-RELATED ILLNESS
Alaimo et al, 2008 [96]	Good										√						↑		
Blair et al, 1991 [95]	Fair								√								↑		
Branas et al, 2009 [61]	Good	√												↑					
Brown & Werner, 2009 (Journal of the APA) [86]	Good					√										↑		↓	
Brown et al 2009 (Health & Place) [76]	Good			√		√												↓ ↓	
Burdette & Whitaker, 2004 [181]	Fair											√						↔	
Carter et al, 2003 [26]	Fair		√											↓					
Coogan et al, 2009 [86]	Good					√										↑			
Dengel et al, 2009 [98]	Fair											√							↑
Dumbaugh & Rae, 2009 [44]	Good				√										↓				
Frank et al, 2004 [71]	Good			√												↑		↓	
Frank et al, 2005 [23]	Good			√												↑			
Gralova, 2008 [93]	Good							√		√	√							↔ ↑ ↔	

Figure 43. Best Practice: Transparent Literature Search/ Review Documentation – a tabular summary of literature review results (Source: HIA of the Transform Baltimore Zoning Code Rewrite).

Weight	Study Type	Measurements	Source
More	Population-based	Measured	Peer-review Journals
	Risk assessment	Validated model	Public health/medical reports
	Case series/ case reports	Non-validated model	Publications by public health authorities
	Animal studies		Publications by other groups (Industry, community members)
Less			Other: Web sites, news articles, opinions, etc.

Figure 44. Best Practice: Quality of Evidence Evaluation – a hierarchy of evidence (Source: HIA on Wind Energy Development in Oregon).

Table 34 Key Data Gaps by Health Effect Category

HEC 1 Social Determinants of Health (SDH)	Current household level survey data for the PACs is not available. Data sets do not include years 2009-2011.
HEC 2: Accidents and Injuries	Data sets do not include data from the ATR for years 2009-2011.
HEC 3 Exposure to potentially Hazardous Materials	No offsite residential monitoring well data are available for review. Fish/Aquatics data set may not fully capture all of the recent restoration efforts. There are no site specific $PM_{2.5}$ data. There is no air permit application available for review There is no analysis of potential dust/diesel emissions in Point MacKenzie. Visual effects analysis is not available. Complex off-Site terrain noise modeling has not been performed.

Figure 45. Best Practice: Identification of Data Gaps – key data gaps by health effect category (Source: HIA for the Proposed Coal Mine at Wishbone Hill).

Variable	Data Source	Geographic Unit of Analysis	HIA Analyses
Transportation and Land Use Conditions: 2005 , 2015 BAU, 2015 RP			
City Lots, Building Heights, Zoning Designation	SF Planning	Lot	AQ, Noise
Pedestrian Environmental Quality	PEQI Data Collection	Street Segment, Intersection	Ped Inj
Traffic Volume by Vehicle Type, Time of Day, Traffic Speed (Free Flow and Congested Conditions), Bus Volumes	SFCTA Model	Street Segment	AQ, Noise, Cyclist Inj, Ped Inj, Equity
Trips, Travel Mode by Age Category	BATS	Bay Area Region	Active Transport
Walk and Bike Trips, Travel Time	SFCTA Model	Transportation District	Active Transport, Cyclist Inj
Socio-demographic Conditions: 2005 and 2015			
Resident Age	AGS Inc. Estimates	Census Block	All
Average Household Income	ABAG Estimates	Transportation Analysis Zone	Equity
Resident Population Data	ABAG Estimates	Transportation Analysis Zone	All
Health Outcomes and Behaviors: 2005 Existing Conditions			
Mortality	CDPH (Vital Statistics)	County	AQ, Active Transport
Myocardial Infarction, Hospital Admission	CDPH (OSHPD)	County	Noise
Vehicle-Pedestrian Injury Collisions, Vehicle Cyclist Injury Collisions	SWITRS	Intersection, Census Tract, County	Ped Inj, Cyclist Inj

Figure 46. Best Practice: Detailed Documentation of Data and Methodology – key input data (Source: Health Effects of Road Pricing in San Francisco HIA).

Indicator	Data Source	Methodology
Number of employees by industry[hh] for all geographies	US Census LEHD Workplace Area Characteristics, 2008	Aggregated total workers in each block by industry for the Central Corridor (CC), Saint Paul, and Ramsey County in SPSS
Average wages by industry for all industries	BLS Quarterly Census of Employment and Wages, 2009	At QCEW web site, downloaded average annual pay by 2-digit NAICS code for all industries, total covered ownerships, all establishment sizes, and all employees in Ramsey County. For Public Administration, took average of the average annual pay for local, state, and federal government
Number of employed residents by industry	US Census LEHD Resident Area Characteristics, 2008	Aggregated total workers in each block in the CC by industry in SPSS
Number of jobs in the CC by place of worker residence	US Census LEHD Origin-Destination Employment Statistics, 2008	In SPSS, selected all blocks with jobs in Ramsey County; from those blocks, selected all blocks with workers in Ramsey County. Coded those blocks to identify those for workers that reside in Saint Paul in the CC
Number of employed residents in the Corridor by place of work	US Census LEHD Origin-Destination Employment Statistics, 2008	In SPSS, selected all blocks with residents who live in the CC. Coded those blocks to identify those who work in the CC, Saint Paul, and Ramsey County
Educational attainment of workers by industry	Census Equal Employment Opportunity data set, 2000	Downloaded Educational Attainment (5 levels) by Census Occupational Codes by worksite for Ramsey County. Downloaded Employment by Census Occupational Codes and Industry by worksite for Ramsey County. Used Access to relate the data sets and calculate the percentage of workers with less than high school education, a high school diploma, some college or an Associate's, Bachelor's, and Master's degree or greater by industry
Educational attainment distribution overall and by race, age	US Census 1990, 2000; American Community Survey 2005–2009 5-year estimates	Aggregated educational attainment for all residents and by race for all block groups in the CC; aggregated educational attainment by age for all census tracts in the CC (2005–2009 data only); downloaded educational attainment overall and by race/age for Saint Paul and Ramsey County on US Census Bureau's American FactFinder

Figure 47. Best Practice: Detailed Documentation of Data and Methodology – a detailed description of data analysis methodology (Source: St. Paul Healthy Corridors for All HIA).

The main potential exposure pathways to contaminants produced by regional oil and gas activities for residents of the region would include:

1) Consumption of tainted subsistence resources: pollutants from oil and gas operations could contaminate local subsistence resources, and expose individuals to contaminants when the harvested resource is consumed.

2) Inhalation: emissions from combustion associated with exploration and production activities could be entrained in the local airshed, and inhaled by residents; subsistence hunters travelling near combustion sites, and residents nearest major emissions sources would be at greatest risk. It is important to recognize that even projects complying with NAAQS standards may produce levels of pollutants that are harmful to human health, particularly vulnerable groups such as infants, elders, and people with underlying chronic illnesses (U.S. EPA 2006; U.S. EPA Region IX 2008; U.S. EPA 2008).

3) Direct contact with skin (as could occur in the case of an oil spill).

4) Contaminated drinking water: Drinking water in the NSB is generally taken from surface water bodies, which could become contaminated through local oil and gas activities.

Figure 48. Best Practice: Impact Pathways/Logic Frameworks – exposure pathways to contaminants (Source: Beaufort Sea and Chukchi Sea Oil and Gas Lease HIA).

Table 8: Summary of expected health impacts from modification of the Trenton Farmers' Market

Pathway ↓	Alternative 1: No-change/minor change		Alternative 2: Full implementation of PPS recommendations: major remodeling		Alternative 3: Market outreach/ satellite markets	
Nutrition (e.g. consumption of fresh fruits and vegetables)	0	Changes to the market too small to significantly impact food access and consumption	0	Patronage and sales may increase, but these changes would probably not change consumption patterns, since there's no indication that changes would affect individuals with poor food access	+	Satellite markets would target neighborhoods and populations with limited access to fresh produce.
Direct Economics Effects (e.g. increased income for vendors)	+/0	Some small increase in patronage and revenues could occur as a result of minor cosmetic changes to facility.	+	Expansion of the market and increasing market activities during the low season, coupled with improved visibility will likely lead to a substantial increase in sales with subsequent increases in income for vendors.	+	Expansion of outlets, broadening of customer base will likely increase sales and income to vendors, but probably not as much as in Alternative 2.
Second-order economic effects (e.g. neighborhood economic expansion and development)	0	Any increase in revenue would be unlikely to be large enough to generate secondary economic impacts.	+	Increased patronage and sales are likely to generate secondary economic benefits through "recycling" of income, by attracting customers to other nearby businesses, and by stimulating neighborhood redevelopment efforts.	0	Modest expansion of sales potential under this option would probably be insufficient to yield second-order economic impacts on the surrounding community.
Physical Activity (e.g. walking and biking to the market)	0	Changes to the market too small to change patterns of physical activity.	+	Redevelopment in surrounding neighborhood could improve walkability/bikeability and induce more people to walk /bike to the market. Improvements in bus service, coupled with outreach to transit-dependent populations could increase walking associated with bus trips to the market.	0	Bringing the market to people would minimize travel distance, thus walking trips to the market would not increase. This alternative by itself would not be sufficient to spur neighborhood redevelopment with improvements in walkability.
Social Capital (e.g. opportunities to socialize with other residents, develop social networks)	0	Changes to the market too small to change community social capital.	+	Increases in market patronage, using market facilities for community meetings and events, and subsequent redevelopment could all contribute to improved community social capital.	+	Could benefit community social capital. Depends on reaching new patrons and providing events that draw residents. May also improve sense of community of it becomes seen as neighborhood asset.
Preventive health services (e.g. health education and screening services on site)	0	No additional preventive services planned under this alternative	+/0	Impacts on preventive health services available at the market contingent on agencies and organizations deciding to bring such services to the market.	+/0	Satellite market at the Capital Health Systems hospital would facilitate tie-in to various health services. Contingent on hospital and health department decisions.

"0" (no change), "+" (potentially beneficial), "-" (potentially harmful)

Figure 49. Best Practice: Clear Summary of Impact Assessment – a summary of differential impacts of analyzed alternatives (Source: Trenton Farmers Market HIA).

Table 1: HIA Analysis Summary of Findings

Health Determinant	Direction	Magnitude	Impact	Significance Likelihood	Distribution
Traffic Safety	↑	High	High	Very Likely	Affects whole community relatively equally
Physical Activity	↑	Medium	Medium/High	Very Likely	Impacts neighboring vulnerable community and whole community via expanded access
Access to Goods and Services	↑	Medium	Medium/High	Very Likely	Disproportional effect on low income, transit-dependent communities around DMA
Air Quality	↑	Low	Low	Possible	Affects whole community relatively equally

Legend:

Direction of Impact:

Positive = Changes that may improve health
Negative = Changes that may detract from health
Uncertain = Unknown how health will be impacted
No effect = No effect on health

Magnitude of Impact: (realizing the proposed project is a 2.1 mile stretch of road, so the comparison or point of perspective is those who currently use DMA)

Low = Causes impacts to no or very few people
Medium = Causes impacts to wider number of people
High = Causes impacts to many people
Note that this is relative to population size

Significance of Impact:

Low = Causes negative impacts that can be quickly and easily managed or do not require treatment or causes positive impacts that are not serious/significant
Medium = Causes negative impacts that necessitate treatment or medical management and are reversible or positive impacts that provide opportunity for improved health
High = Causes or prevents death or serious illness

Likelihood of Impact:

Very Likely = it is very likely that impacts will occur as a result of the proposal
Likely = it is likely that impacts will occur as a result of the proposal
Possible = it is possible that impacts will occur as a result of the proposal
Unlikely = it is unlikely that impacts will occur as a result of the proposal
Uncertain = it is unclear if impacts will occur as a result of the proposal

Distribution of Impact:

The community surrounding DMA has a large minority population (37.4% in the 29303 zip code and 70.8% of the population in the 29306 zip code). When it comes to economic characteristics, in the 29303 zip code 18.1% of individuals live below the poverty line and in 29306 28.8% of individuals live below the poverty line.

Figure 50. Best Practice: Clear Summary of Impact Assessment – a summary of health impact assessment findings (Source: Daniel Morgan Road Diet and Restriping HIA).

Uncertainty Factors Regarding the Magnitude of Estimated Health Effects for Lives Saved from Active Transportation via Walking and Cycling

Factors Affecting Certainty	Assessment Approach	Summary Confidence Level
Exposure Assessment	Changes in walking and cycling trips and time for transportation based on model outputs from the SFCTA model at a district level, the smallest area level with reliable estimates. Used BATS (2000) data to estimate trips by age, assuming the proportion of the population travelling during the average weekday and average number of daily trips are the same within San Francisco as the rest of the Bay Area. Does not include leisure walking and cycling trips, or walking trips to transit. Likely an underestimate of walking and cycling, overall.	Moderate
Baseline Disease Prevalence	Mortality Rate: County-level data from vital statistics for people aged 25-64.	High
Exposure-Response Function (ERF)	Health Economic Assessment Tool (HEAT) for walking and cycling approach. Estimate for walking based on meta-analysis of nine studies. Estimate for cycling based on three studies and expert consensus. A threshold approach, though evidence for an inverse-linear relationship, which could result in an underestimate of overall benefits. Overall health benefits underestimated given tool focus on adults, mortality only.	Moderate

Figure 51. Best Practice: Confidence Estimates/Assessments of Uncertainty–characterization of uncertainty in assessment (Source: Health Effects of Road Pricing in San Francisco HIA).

Parameter Ranking for Prioritizing HECs

Rank	Stakeholder Concern	Data Gaps	Potential Impact	Likelihood
Low	HEC not mentioned by stakeholders	No data gaps exist for the HEC	HEC review reveals no potential impacts or minor positive or negative impacts only; easy adaptation	Unlikely: HEC effects, not typically observed in similar settings or have never been observed in similar settings
Medium	HEC mentioned by stakeholders; somewhat important	Data gaps exist for the HEC but would provide little additional information if filled	Modest increase (or decrease) in mild to moderate disease events; effective mitigation strategies exist	Possible: At least some of the health effects in this HEC have occurred in similar settings and could occur given proper conditions or design features
High	HEC highly important to stakeholders	Data gaps exist for the HEC and represent critically important information	Substantial increase (or decrease) in moderate to severe morbidity or mortality events; adaptation with assistance	Probable: At least some of the health effects in this HEC are known to be common in similar settings and will most likely occur

Figure 52. Best Practice: Prioritization Process for Recommendations – qualitative ranking system to prioritize health effect categories for action (Source: HIA for the Proposed Coal Mine at Wishbone Hill).

EMPLOYMENT

Strategy	Recommendations	Decision-maker	Source of Recommendation	Relation to Top Recommendations
Policy	1. Recruit businesses that prioritize employing local residents.	Beyond Housing; City of Pagedale	Community	
	2. Incentivize Pagedale employers to offer a Living Wage at a minimum of $8.30/hr.	City of Pagedale	Research	
	3. Incentivize employment of residents in new retail and to fund workforce development programs with a development agreement or community benefits agreement.	City of Pagedale; Developers	Research	
Design	4. Design buildings for mixed use.	Beyond Housing	Research	
	5. Provide space for a community marketplace to support local microenterprises.	City of Pagedale; UMSL Extension	Key Informant	
Program	6. Increase entrepreneurship training and business management support to job training services to improve the employability of residents.	Social service agencies	Key Informant	3
Education	7. Inform residents of potential job opportunities offered by new enterprises.	Social service agencies	Community	3
	8. Provide information about job training and job application.	Social service agencies	Community	3

Figure 53. Best Practice: Implementation Plan for Recommendations – recommendation strategies and decision-makers (Source: Page Avenue HIA)

RECOMMENDATIONS FOR A MIXED USE, MIXED INCOME COMMUNITY

Health Promoting	Potential Health Impacts	Equity Issues	Summary	Design Indicators
Access to Public Transit	**Transit Benefits for Individuals**: Almost one-third of Americans who commute to work via public transit meet their daily requirements for physical activity (30 or more minutes per day) by walking as part of their daily life, including to and from the transit stop (Besser & Dannenberg, 2005) **Transit Use by Proximity**: Proximity to public transit helps determine travel choices (Ewing 2006). For every ¼ mile increase in distance to public transit from homes, there was an associated 16% decrease in transit use. (Lawrence Frank & Company, 2005) **Transit Benefit and Environment**: Increased public transit use directly results in decreased air pollution from passenger vehicles (Ewing 2006).	Currently 70% of sub-area residents live within 1/4 mile of transit stop. Less served areas are predominantly lighter residential and **higher income**. Access to transit by **elderly** needs to be facilitated by ensuring close proximity to transit stops.	*If achieved:* ▲ Physical Activity ▲ Use of transit ▲ Health equity *If not achieved:* Status Quo	Transit Routes. Access to transit stops. Access to employment, goods, and services.
Affordable Housing	**Benefits of Affordable Housing**: More equitable, affordable housing increases social cohesion, decreases displacement and homelessness, decreases stress, increases overall health (SFDPH 2004). The National Low Income Housing Coalition classifies people as severely cost-burdened if more than 50% of their household's income is spent on housing costs (theHDMT.org). Lack of affordable housing can increase poverty, crowding, displacement, and homelessness all of which result in poor health outcomes (SFDPH 2004; Hood, 2005). **Benefits of Relocation:** Relocating residents from public housing projects into neighborhoods with lower concentrations of poverty has been associated with weight loss and a decline in reported stress levels among adults, and reduced rates of injury among male youths (Orr et al, 2003; SFDPH, 2004).	**Gentrification** can push out **low-income** and **ethnic minority** residents, whose housing stability and health outcomes will be affected adversely by increases in housing prices. Housing planning can also result in segregation, which has adverse impacts for health of **African Americans**. In the Highway 99 area, **Hispanics** would be the largest ethnic group and most likely to be affected.	*If achieved:* ▲ Social Cohesion ▲ Social Equity ▼ Stress ▼ Homelessness ▼ Obesity ▼ Injury *If not achieved:* ▲ Poverty ▲ Crowding ▲ Homelessness ▲ Cost burden	Stability of housing values Availability of affordable housing in proportion to local household incomes and demand Displacement rates Relocation planning and assistance Mixed age, ethnicity, income levels

Figure 54. Best Practice: Implementation Plan for Recommendations – recommendation summary and indicators (Source: Highway 99 Sub-Area Plan HIA).

Stage 5: Generating *Measurable Indicators and Element Profiles* (April 2005—August 2005)

SUMMARY:

In this stage, the ENCHIA process worked to identify indicators of community health that could help measure how well the City was performing with respect to the ENCHIA *Healthy City Vision* and *Community Health Objectives*. Council subgroups and ENCHIA staff gathered a significant amount of data to generate baseline data profiles. Staff also completed qualitative research on how development was impacting specific population subgroups underrepresented on the Community Council as well as on the relationship of health to psycho-social employment attributes.

KEY OUTCOMES/ACHIEVEMENTS:

- Gathered data on over 100 measurable community health indicators
- Generated five *Element Profiles* using quantitative data on selected indicators
- Completed study and report titled: *Tales of the City's Workers: A Work and Health Survey of San Francisco's Workforce*
- Completed study and report titled: *Eastern Neighborhood Community Health Impact Assessment: Results from a Community Assessment of Health and Land Use*

TASKS/ACTIVITIES:

- Conducted research on characteristics of good indicators
- Subgroups collected and reviewed baseline data on selected indicators
- Disaggregated indicators by variables of interest such as race/ethnicity, income, and geography
- Presented data to the larger Council to get feedback on selected indicators and ideas for new indicators
- Conducted numerous focus groups and key informant interviews to complete qualitative studies

Figure 55. Best Practice: Clear/Transparent HIA Report – transparent documentation of a stage of the HIA process (Source: Eastern Neighborhoods Community HIA).

In the case of RAD, we propose the following monitoring plan:

1. Monitoring the impact of this HIA on the decision: National People's Action will be responsible for tracking the progress of RAD to monitor if RAD has been amended according to this HIA's recommendations, and whether this HIA had any influence on the thinking of policy-makers in terms of the evaluation of RAD and the expansion of RAD beyond the pilot period into a permanent policy.

2. Monitoring decision implementation: RAD's evaluation process should include a Conversion Oversight Committee made up of resident organizations, public housing advocates, and elected officials to monitor RAD's implementation. Their monitoring will include tracking selection criteria for public housing complexes chosen for conversion, how decisions about selection are made, and allocation of funding for conversion, relocation counseling, and other programming or support services related to policy implementation. This information shall be reported out semi-annually for two years. If a Conversion Oversight Committee is not created, NPA and Advancement Project will try to work with HUD to report out these indicators semi-annually for two years.

3. Monitoring determinants of health: If RAD's evaluation process includes a Conversion Oversight Committee, the Committee will obtain information from HUD regarding the number of housing complexes and number of units "converted" from HUD ownership to another entity's ownership; the number of units that remain available for very low- and low-income residents; the number of management systems changed from public to private management; the number of any lost units; the number of evictions; and, the number of vouchers created and used. Impacts on health will be assessed via impacts on these changes. Longer-term impacts on health and tracking of impacts on residents will be monitored, ideally, via HUD evaluation of RAD implementation, or pending further funding.

Figure 56. Best Practice: Monitoring Plan –impact and outcome evaluation (Source: Rental Assistance Demonstration Project HIA).

Areas for Improvement

There are ample areas for improvement in the HIA community of practice. Addressing these will help to improve the overall quality and rigor of HIAs, advance the community of practice, and improve overall utilization and effectiveness of HIAs in the U.S. Areas for improvement include: adherence to the *Minimum Elements and Practice Standards*, use of HIA to inform decision-making at all levels, consistency in terminology, broader utilization of existing tools and resources, and identification and closing of data gaps.

Adherence to *Minimum Elements and Practice Standards*

Adherence to the *Minimum Elements and Practice Standards for Health Impact Assessment* (North American HIA Practice Standards Working Group 2010) or similar criteria developed by the National Research Council (2011) is undoubtedly the single most important factor for advancing HIAs in the U.S. The *Practice Standards* provide HIA benchmarks or best practices that HIA practitioners should strive to meet when performing HIAs. If these standards were implemented more consistently, the results would be extraordinary. Consistent implementation of the *Minimum Elements* (or National Research Council criteria) across the HIA community of practice would ensure the essential components of HIA are put into practice that help distinguish HIA from other practices and methods, and would result in marked increases in rigor, quality, defensibility, and effectiveness. Essential components of HIA that are particularly lacking and should be targeted for improvement are given below.

Establishment of Baseline Conditions
A profile of baseline conditions is necessary in order to identify disparities in existing conditions, predict future conditions if the decision is implemented, and compare with future conditions should the decision be enacted (i.e., for impact monitoring).

Characterization of Impacts
Consistency in judging direction, magnitude, likelihood, distribution, and permanence of impacts and increased use of quantitative evaluation methods, when warranted, would increase the defensibility and effectiveness of HIA considerably. Due to the current lack of consideration in impact assessment of likelihood, magnitude, permanence, and quantification, further guidance or methodology may be needed for the HIA field of practice to reach a state of consistency in this area.

Stakeholder and Community Engagement
Stakeholder engagement is central to the HIA field of practice, yet it is lacking in many HIAs. Enforcing the need for deliberative engagement of the community, decision-makers, and other stakeholders in the HIA process will lead to greater depth in the evidence base, promote equity, empower communities, and ensure that the recommendations offered are relevant to the community and feasible, practical, and able to be implemented within the purview and authority of the decision-makers.

Transparency in Documentation

Effective communication of HIA results is crucial to informing the decision-making process, and effective communication of HIA processes, strategies, methodologies, and lessons learned (including some components of the HIA process that may be documented separate from the traditional HIA Report, such as communication plans and monitoring and evaluation efforts) are crucial for advancing the HIA community of practice; both of these can best be achieved through clarity and transparency in developed reporting materials. Following the guidance provided in the *Minimum Elements and Practice Standards* would result in more consistent transparency and quality in HIA documentation.

Monitoring and Evaluation

Monitoring and evaluation is an area that is considerably lacking in the HIA practice. Monitoring (i.e., impact evaluation and outcome evaluation) is crucial for improved utilization and effectiveness of HIAs because it establishes baselines and trends for accountability, builds a better understanding of the value of HIA, and validates the accuracy of health impact predictions. Further research should be conducted on developing strategies and methodologies to minimize the challenges that exist in implementing monitoring measures.

While monitoring can be challenging to implement, process evaluation is an element that could be easily incorporated into the HIA process. Process evaluation can be used not only to document the defensibility of a conducted HIA, but also results in valuable information (e.g., successes, challenges, lessons learned, etc.) that can be used to help refine methods and approaches used in HIA and advance the HIA community of practice.

In addition to the value of implementing monitoring and evaluation for individual HIAs, there is considerable potential for these efforts to advance the HIA community of practice. However, the value of monitoring and evaluation for the practice as a whole will not be realized if the documentation of those efforts (e.g., plans for and results of monitoring and evaluation) are kept internally and not made available to the community of HIA practitioners.

Use of HIA to Inform Decision-making at All Levels

HIA is commonly implemented to inform local decisions; however, strategies should be developed for applying HIA more readily to decisions at all levels, including county, state, and federal decisions.

Consistency in HIA Terminology

Inconsistencies exist in the sector terminology used by the organizations utilized in the preliminary literature search for the HIA Review. Figure 57 shows the inconsistencies in sector terminology among these various organizations, which are responsible for promoting and/or reporting HIAs. During the HIA Review, inconsistencies were also noted in pathway

102

terminology. Like transparency, consistency in terminology will help to advance HIA reporting and rigor.

HEALTH**IMPACT** P R O J E C T — The HIA Gateway		World Health Organization	UCLA HEALTH IMPACT ASSESSMENT CLEARINGHOUSE LEARNING & INFORMATION CENTER
Agriculture and Food	Agricultural Industry	Agriculture	Agriculture
Built Environment	Air Transport	Air	Communications, Media
Climate Change	Crime	Culture	Community Planning
Economic Policy	Economic Policy	Development	Economic Policy
Education	Education	Energy	Education
Gambling	Employment	Environment	Environmental Protection
Housing	Greenspace	Housing	Food Processing, Distribution, Sales
Labor and Employment	Health Services	Integrated impact assessment	Housing
Natural Resources and Energy	Housing	Mining	Labor, Workplace
Physical Activity	Income	Noise	Land-Use Planning
Transportation	Industry	Other subjects	Mining, petroleum, other extractive industry
Water	Inequalities	Overview	Parks & Recreation
	Leisure	Recreation	Public Safety
	Planning	Social welfare	Taxation
	Regeneration	Tourism	Transportation
	Roads	Transport and	Utilities
	Sustainable development	communications	Workplace
	Transport	Waste	
	Urban Areas	Water	
	Waste		

Figure 57. Inconsistencies in sector terminology among various organizations promoting and/or reporting HIAs.

Broader Utilization of Existing Tools and Resources in HIAs

Broader utilization of existing tools and resources could contribute to a more robust impact assessment and help to close some of the data gaps found in HIA. A comprehensive inventory of tools, models, and methodologies that can be used in HIA would greatly benefit the HIA community of practice. An effort is currently underway to develop an HIA Roadmap for incorporation into the Community-Focused Exposure and Risk Screening Tool (C-FERST) under development by the EPA. This HIA Roadmap is envisioned to include an inventory of models, tools, and other resources for use in HIA. Table 12 highlights C-FERST and a few other tools and models not utilized in the reviewed HIAs, but that could be of benefit in the HIA practice.

Table 12. Additional Tools and Models Useful for the HIA Community of Practice

Tool/Model	Description	Source
AirData	Provides access to the EPA's Air Quality System (AQS) Data Mart which is updated each week night with air quality data collected at outdoor monitors across the U.S. One can get criteria pollutant and air quality index (AQI) data in multiple forms, including reports, graphs, maps, and other visualization forms.	Environmental Protection Agency; http://www.epa.gov/airdata/
Benefit Mapping and Analysis Program (BenMAP)	A GIS-based computer program used to estimate the health impacts and associated economic value experienced with changes in air quality.	Environmental Protection Agency; http://www.epa.gov/air/benmap
Co-Benefits Risk Assessment (COBRA)	A free tool that estimates the health and economic benefits of air quality policies. Allows users to estimate and map the air quality, human health, and related economic benefits (excluding energy cost savings) of clean energy policies or programs; and approximate the outcomes of clean energy policies that change emissions of particulate matter ($PM_{2.5}$), sulfur dioxide (SO_2), nitrogen oxides (NO_X), ammonia (NH_3), and volatile organic compounds (VOCs) at the county, state, regional, or national level.	Environmental Protection Agency; http://www.epa.gov/statelocalclimate/resources/cobra.html
Community-Focused Exposure and Risk Screening Tool (C-FERST)	A one-stop community mapping, information access, and assessment tool designed to help assess risk and assist in decision making within communities. The addition of an HIA Roadmap to this tool will make C-FERST a very useful resource for those that are new to HIA.	Environmental Protection Agency; http://www.epa.gov/heasd/c-ferst
Comparative Quantification of Health Risks	Quantifies risk factor exposure and effects for 26 major health risks and identifies population exposure distributions, evidence for causality, and estimates of disease-specific hazards associated with each level of exposure.	World Health Organization; http://www.who.int/healthinfo/global_burden_disease/cra/en/
Eco-Health Relationship Browser	Illustrates the linkages between human health and ecosystem services (i.e., benefits supplied by nature). This tool provides information about several major ecosystems in the U.S. (i.e., agro-ecosystems, forests, urban ecosystems, and wetlands), the services they provide, and how those services, or their degradation and loss, may affect people.	Environmental Protection Agency; http://www.epa.gov/research/healthscience/browser/introduction.html
MyEnvironment	Integrates data from EPA and other sources, including air, water, energy, and health data, to provide a quick picture of local environmental conditions.	Environmental Protection Agency; http://www.epa.gov/myenvironment
ParkScore	Provides measures of how well the 40 largest U.S. cities meet their need for parks. Using advanced GIS, ParkScore identifies neighborhoods and demographics that are underserved by parks, and the number of people able to reach a park within a ten-minute walk. It also provides in-depth data to guide local park improvement efforts.	Trust for Public Land; http://parkscore.tpl.org/
Regional Vulnerability Assessment (REVA) Environmental Decision Toolkit	A web-based application that provides a means for visualizing and exploring data about current conditions, possible future conditions, and integrating stressors and conditions. The objective of ReVA is to assist decision-makers in making more informed decisions and in estimating the large-scale changes that might result from their actions.	Environmental Protection Agency; http://amethyst.epa.gov/revatoolkit/Welcome.jsp

Identifying and Closing the Data Gaps in HIA

Identification of data gaps is important to transparency in HIA reporting, but it can also be useful in helping to refine methods and approaches used in HIA and identify areas for future research. Closing identified data gaps and maximizing the evidence available for use in HIA will result in more robust assessments and improved efficiency in predicting health impacts.

Conclusions

While HIAs have helped to raise awareness and bring health into decisions outside traditional health-related fields, the effectiveness of HIAs in bringing health-related changes to pending decisions in the U.S. varies greatly. This review found that there are considerable disparities in the quality and rigor of HIAs being conducted. This, combined with lack of monitoring, health impact management, and other follow-up in HIAs could be limiting the overall utilization and effectiveness of this tool in the U.S.

Understanding the current state and applicability of HIAs in the U.S., as well as best practices and areas for improvement, will help to advance the HIA community of practice in the U.S., improve the quality of assessments upon which stakeholder and policy decisions are based, and promote healthy and sustainable communities.

References

AAG. 2012. Annals of the Association of American Geographers Special Issue: Geographies of Health. Philadelphia, PA: Taylor & Francis Group.

Alaska Department of Health and Human Services. 2011. Technical Guidance for Health Impact Assessment (HIA) in Alaska. Juneau, AK: Alaska Department of Health and Human Services.

Arnstein, S.R. 1969. A ladder of citizen participation. Journal of the American Institute of Planners 35(4):216-224.

Bhatia, R. 2011. Health Impact Assessment: A Guide for Practice. Oakland, CA: Human Impact Partners.

Bhatia, R. and M. Katz. 2001. Estimation of health benefits from a local living wage ordinance. American Journal of Public Health 91:1398-1402.

Bhatia, R. and A. Wernham. 2008. Integrating human health into environmental impact assessment: an unrealized opportunity for environmental health and justice. Environmental Health Perspectives 116(8):991-1000.

Danneberg, A.L., R. Bhatia, B.L. Cole, S.K. Heaton, J.D. Feldman, and C.D. Rutt. 2008. Use of health impact assessment in the U.S.: 27 case studies, 1999-2007. American Journal of Preventive Medicine 34(3):241-256.

Deganian, D., and J. Thompson. 2012. Patterns of Pollution: A Report on Demographics and Pollution in Metro Atlanta. Atlanta, GA: GreenLaw.

Design for Health. 2008. Rapid Health Impact Assessment Toolkit, Version 3.0. Minneapolis, MN: University of Minnesota.

EPA. 2011. Draft Research Framework: Sustainable and Healthy Communities Research Program. Washington, DC: U.S. Environmental Protection Agency.

Gotway, C.A., and L.J. Young. 2002. Combining incompatible spatial data. Journal of the American Statistical Association 97(458):632-648.

Harris, P., B. Harris-Roxas, E. Harris, and L. Kemp. 2007. Health Impact Assessment: A Practical Guide. Sydney: Centre for Health Equity Training, Research and Evaluation (CHETRE).

Harris-Roxas, B. and E. Harris. 2011. Differing forms, differing purposes: A typology of health impact assessment. Environmental Impact Assessment Review 31(4):396-403.

Health Impact Project. 2013. The Rise of HIAs in the U.S. Washington, DC: Pew Charitable Trusts. Available at: http://www.pewhealth.org/reports-analysis/data-visualizations/the-rise-of-hias-in-the-united-states-85899464695.

Heller, J., S. Malekafzali, L.C. Todman, and M. Wier. 2013. Promoting Equity through the Practice of Health Impact Assessment. Oakland, CA: PolicyLink

Human Impact Partners. 2011. A Health Impact Assessment Toolkit: A Handbook to Conducting HIA, 3rd edition. Oakland, CA: Human Impact Partners.

Human Impact Partners. 2012. HIA Summary Guides. Oakland, CA: Human Impact Partners.

ICMM. 2010. Good Practice Guidance on Health Impact Assessment. London, UK: International Council on Mining and Metals.

IPIECA/OGP. 2000. A Guide to Health Impact Assessments in the Oil and Gas Industry. London, UK: International Petroleum Industry Environmental Conservation Association/ International Association of Oil & Gas Producers.

National Research Council. 2011. Improving Health in the United States: The Role of Health Impact Assessment. Washington, D.C.: The National Academies Press.

North American HIA Practice Standards Working Group (Bhatia, R., J. Branscomb, L. Farhang, M. Lee, M. Orenstein, and M. Richardson). 2010. Minimum Elements and Practice Standards for Health Impact Assessment, Version 2. Oakland, CA: North American HIA Practice Standards Working Group.

Quigley, R., L. den Broeder, P. Furu, A. Bond, B. Cave, and R. Bos. 2006. Health Impact Assessment International Best Practice Principles. Special Publication Series No. 5. Fargo, ND: International Association for Impact Assessment.

SFDPH. 2011. Health Effects of Road Pricing In San Francisco, California: Findings from a Health Impact Assessment. San Francisco, CA: San Francisco Department of Public Health, Program on Health, Equity, and Sustainability.

Stakeholder Participation Working Group. 2010. Best Practices for Stakeholder Participation in Health Impact Assessment, Version 1.0. Oakland, CA: Stakeholder Participation Working Group of the 2010 HIA in the Americas Workshop.

UCBHIG. 2009. HOPE VI to HOPE SF San Francisco Public Housing Redevelopment: A Health Impact Assessment. Berkeley, CA: University of California Berkeley Health Impact Group.

Warburton, D., L. Colbourne, K. Gavelin, and R. Wilson. 2008. Deliberative Public Engagement: Nine Principals. London: National Consumer Council.

WHO. 1999. Health Impact Assessment: Main Concepts and Suggested Approach; Gothenburg Consensus Paper. Copenhagen, Denmark: World Health Organization, Regional Office for Europe.

Winkler, M.S., M.J. Divall, G.R. Krieger, M.Z. Balge, B.H. Singer, and J.Utzinger. 2010. Assessing health impacts in complex eco-epidemiological settings in the humid tropics: Advancing tools and methods. Environmental Impact Assessment Review 30(1):52-61.

Wismar, M., J. Blau, K. Ernst, and J. Figueras (Eds.). 2007. The Effectiveness of Health Impact Assessment: Scope and Limitations of Supporting Decision-making in Europe. Copenhagen, Denmark: World Health Organization.

World Bank Group. 2006. IFC Sustainability Framework. Washington, D.C.: World Bank Group, International Finance Corporation.

Young, L.J., C.A. Gotway, J. Yang, G. Kearney, C. DuClos. 2009. Linking health and environmental data in geographical analysis: It's so much more than centroids. Spatial and Spatio-temporal Epidemiology 1:73–84.

HEALTH IMPACT ASSESSMENT REVIEW GUIDELINES

May 30, 2012
(Annotated July 2012, August 2013)

Compiled By: Justicia Rhodus, Dynamac Corporation

HEALTH IMPACT ASSESSMENT REVIEW
BACKGROUND

SCOPE

A systematic review is being conducted of health impact assessments (HIAs) from the U.S. to obtain a clear picture of how HIAs are being implemented nationally and to identify potential areas for improving the HIA community of practice. The review is focused on HIAs from four sectors that the U.S. Environmental Protection Agency's (EPA's) Sustainable and Healthy Communities Research Program has identified as targets for empowering communities to move toward more sustainable states – transportation, housing/buildings/infrastructure, land use, and waste management/revitalization.

A pilot review of four HIAs – one from each sector – has been completed to date. Through this process, the project team has been able to refine the information to be recorded from the HIA reviews.

The hope is that the HIA Review will aid in identifying EPA products and research that could enhance the HIA community of practice and discovering what the ecological assessment community of practice could draw from HIAs (and vice versa).

PRODUCTS

The products of this HIA Review will be two-fold:

- an Access database documenting the review of available HIAs in the four identified sectors; and
- a report synthesizing the results of the review to identify the state of the HIA practice in the U.S. and areas in the overall HIA process that could benefit from enhanced guidance, strategies, and methods.

TIMELINE

The HIA reviews are expected to be complete in early August 2012, with release of the report to follow in September 2012.

> * NOTE (July 2012): The timeline for completion of the review has been delayed. It is expected that the report will be issued mid-late 2013.

HEALTH IMPACT ASSESSMENT REVIEW
WHAT IS HIA?

HIA DEFINED

Health impact assessment (HIA) is commonly defined as:

> a combination of procedures, methods, and tools which systematically judges the potential, and sometimes unintended, effects of a policy, program, or project on the health of a population and the distribution of those effects within the population.

---

* NOTE (August 2013): This definition is from the International Association of Impact Assessment (Quigley et al. 2006); an updated definition was developed by the National Research Council in 2011.

Health impact assessment
- **identifies and evaluates public health consequences** of a plan, project, or policy,
- suggests actions to **minimize adverse health impacts and optimize beneficial ones**, and
- provides **recommendations intended to shape the final proposal.**

Health impact assessment is:

- a way to **factor health considerations into the decision-making process**
- **a structured process** that uses scientific data, professional expertise, and stakeholder input
- conducted and communicated **in advance of a decision**
- **both a health protection and health promotion tool** that identifies health hazards and health benefits

HIA STEPS

The major steps in conducting an HIA* include:

1. **Screening** – identify projects or policies for which an HIA would be useful
2. **Scoping** – identify which health effects to consider and setting the parameters of the HIA
3. **Identification** – collect information to identify potential health impacts
4. **Assessment** – synthesize and critically assess the information to prioritize health impacts, identify which people may be affected, and how they may be affected
5. **Decision Making/Recommendations** – make decisions to reach a set of final action-oriented recommendations that promote positive health effects and/or mitigate adverse health effects; write a final report to present the results of the HIA to decision-makers for implementation/action.
6. **Evaluation and Follow-up** – evaluate the processes involved in the HIA, including the effect/impact of the HIA, and set up impact monitoring and a health impact management plan.

---

* NOTE (July 2012): The steps outlined here and noted in the figure on page A-5 and the table on pages A-7 and A-8 were taken from Harris, Harris-Roxas, Harris, and Kemp (2007). It should be noted that the standardized steps for HIAs conducted in the U.S. vary slightly from these and include: Screening, Scoping, Assessment, Recommendations, Reporting, and Monitoring and Evaluation.

HEALTH IMPACT ASSESSMENT REVIEW
HIA LITERATURE SEARCH

PRELIMINARY LITERATURE SEARCH

Health Impact Assessment (HIA) Title	Location	Status	SHCRP Sector(s)	Health Impact Project	HIA Gateway	UCLA HIA Clearinghouse	Human Impact Partners	Other
6th Avenue East Duluth HIA	Duluth, MN	Completed		Built Environment				
A Health Impact Assessment of Mixed Use Redevelopment Nodes and Corridors in Lincoln, Nebraska	Lincoln, NE	Completed			Planning/Transport			
Arctic Outer Continental Shelf Oil and Gas Multiple Lease Sale Environmental Impact Statement	Alaska	Completed		Natural Resources and Energy	Industry			
Atlanta Beltline	Atlanta, GA	Completed		Built Environment	Regeneration	Land Use Planning		
Baltimore Comprehensive Zoning Code Rewrite	Baltimore, MD	Completed		Built Environment		Land Use Planning		
Baltimore Red Line Transit Project	Baltimore, MD	Completed		Transportation		Land Use Planning/Transportation		
Barrett Park Property	Hood River, OR	Completed						http://public.health.oregon.gov/
Battlement Mesa	Battlement Mesa, CO	Completed		Natural Resources and Energy		Mining, Petro, Other Extractive Industry		
Benton Accessory Dwelling Units	Benton County, OR	Completed		Housing				
Bernal Heights Preschool	San Francisco, CA	Completed		Built Environment				
Buford Highway and NE Plaza Redevelopment	Atlanta, GA	Completed		Transportation	Transport	Transportation		
Child Health Impact Assessment of Energy Costs and the Low Income Home Energy Assistance Program (LIHEAP)	Massachusetts	Completed		Natural Resources and Energy		Utilities		
Child Health Impact Assessment of the Massachusetts Rental Voucher Program	Massachusetts	Completed		Housing	Housing/Inequalities	Housing		

For planning purposes, it was estimated in late March 2012 that there were between 80 and 150 HIAs to review in the four identified sectors; this includes the four HIAs in the pilot review: 78 Completed ~ 66 In Progress ~ 3 Undetermined ~ 2 Draft

LITERATURE SEARCH RESULTS

The literature search results were updated in May 2012 and available HIA reports downloaded for review. A total of 89 completed HIAs were obtained.

HEALTH **IMPACT** PROJECT

http://www.healthimpactproject.org

HￍP HUMAN IMPACT PARTNERS

http://www.humanimpact.org/

≈ The HIA Gateway ≈

http://www.hiagateway.org.uk/

UCLA HEALTH IMPACT ASSESSMENT CLEARINGHOUSE
LEARNING & INFORMATION CENTER

http://www.hiaguide.org/

World Health Organization

http://www.who.int/hia/en/

A-4

HEALTH IMPACT ASSESSMENT REVIEW
REVIEW PROCESS

DOCUMENTATION OF REVIEWS

An Access database has been created to record the results of the systematic HIA reviews. An Access 2007 database containing a Data Entry Form will be transmitted to each reviewer. This form, which allows data entry directly into the HIA Review database, will be used to record the results of your HIA Review.

HIA REVIEW

Each reviewer will be assigned a set of HIAs to review. The HIA reports are located on the network at: L:\Public\NERL-PUB\Health Impact Assessment\HIA Review Materials\HIA Reports

Review each assigned HIA report for the information requested in the Data Entry Form shown below. Database fields are defined in more detail on pages 6–7 and expanded upon even further in subsequent pages.

Data Entry Form	
HIA Review	
ID:	Local Data Available or Obtained?:
Title:	Additional Data Needed (Self-identified):
Year:	
Location:	Stakeholder Involvement:
Decision-making Level:	Impacts/Endpoints:
Organization(s) Involved:	Health Endpoints:
Organization Type:	
Contact:	Pathway of Impact:
Organization/HIA Website:	
Funding:	Quantification of Impact:
Status:	Impact Prioritization:
Sector(s):	Decision-making Outcome:
HIA Type:	HIA Report:
HIA Rigor:	Defensibility/Process Eval:
Scope/Summary:	Effectiveness of HIA:
	Follow-up Measures:
	Minimum Elements of HIA Met? If no, what's missing:
Source of Evidence:	GIS Used?:
Data Types:	Environmental/Ecosystem Impacts Considered?:
Major Data Sources:	Potential Improvements:
	Best Practices:

HIA Step { Scoping, Identification | Assessment, Decision/Recommend | Evaluation/ Follow-up

A-5

Data entry should be based on your independent review of the HIA Report and the guidance provided in this document. Some of the requested data will not be directly available in the reports, however – they either require your assessment of the HIA to some extent (HIA Rigor, Defensibility/Process Evaluation, Minimum Elements of HIA Met, Potential Improvements, Best Practices) or request ancillary data, which can be obtained (if available) via an internet search (Organization Type, Effectiveness of HIA).

Please note that some fields have specific terminology or formatting that should be used for data entry; these are highlighted in gray in the database field descriptions on pages 6–7.

Note: If you are having trouble gathering some of the data from the HIA Report, the HIA databases/ clearinghouses identified at the bottom of page 3 can be referenced to gather some of the data (e.g., point of contact, funding, websites, rigor, summary, recommendations, effectiveness of HIA, etc.), although data entries should be based on your review of the HIA to the extent possible*.

* HIA guidelines point to having transparent, publicly-accessible documentation. So, for instance, if you cannot discern the funding sources from the HIA Report, but this information is included in the clearinghouses/databases, do not include it in the HIA Review database; instead, enter "Undetermined." (Note: In the Potential Improvements field, you could enter improved transparency in documentation.)

Likewise, if the information you gather from the HIA Report contradicts information contained in the clearinghouses/databases, enter the information based on your review of the HIA Report and the guidance included here, noting the discrepancy. See this data entry example, noting the discrepancy in HIA Rigor: Intermediate (Listed by Human Impact Project as rapid, but more than three impacts assessed).

BACKGROUND MATERIALS

Materials referenced throughout these guidelines are located on the network at:
L:\Public\NERL-PUB\Health Impact Assessment\HIA Review Materials\Background Materials

HEALTH IMPACT ASSESSMENT REVIEW
DATABASE FIELDS*

These headings are placed in the overall context of the HIA Process (screening, scoping, identification, assessment, decision making and recommendations, evaluation and follow-up), as applicable. The table cells highlighted in gray give specific terms and/or format that must be used for data entry.

Database Field	Description/Examples	HIA Step	Addtl Info on Page(s)
ID	(Automatically generated in Access database)		
Title	Full title of HIA Report		
Year	Year of publication		
Location	Where HIA was conducted– city, county, state, etc. (as applicable)		
Decision-making Level	Local, county, state, federal		
Organization(s) Involved	Organizations involved in conducting/publishing/ sponsoring the HIA		
Organization Type	Educational institution, Government agency, Non-profit, Other, Undetermined		
Contact	Name and contact info for HIA point-of-contact (if available) in format: name, email / Undetermined		
Organization/HIA Website	Identify website dedicated to or highlighting the HIA (if applicable) / N/A		
Funding	Identify financial sponsors (if named) / Undetermined		
Status	Complete, In progress, etc.		
Sector(s)	Transportation, housing/buildings/infrastructure, land use, waste management/revitalization (as defined by SHCRP)		8-12
HIA Type	Mandated (by what/whom), decision support, advocacy, community-led1 [1]		13
HIA Rigor	Desk-based, rapid, intermediate, comprehensive [2]	Scoping	14
Scope/Summary	Question/problem faced, proposed policy/plan examined	Scoping	
Source of Evidence	Literature review, ~~survey,~~ community consultation, policy review, special collection (interviews, surveys, focus groups, risk assessment, demographics analysis, modeling, etc.) [2]	Scoping	
Data Types	Models, literature (published, peer-reviewed, grey lit, government documents, policy), websites, data	Scoping/Identification	15
Major Data Sources	Specific models, agency (e.g., CDC, HUD, Census Bureau*) or community data, bibliographic resources (Medline, Pub Med, Web of Science, Science Direct, etc.), databases, websites, internet gateways/search engines (e.g., Google), surveys, focus groups/forums, entities interviewed/consulted (e.g., stakeholders, technical experts), etc. * **Note:** It has been brought to our attention that it would be useful to note the type, year, and geographic scale of census (and other) data used.	Identification	15
Local Data Available or Obtained?	(If yes) Identify data / No	Identification	
Additional Data Needed (Self-Identified)	(If yes) Identify data / No	Identification	
Stakeholder/Community Involvement?	(If yes) Identify stakeholder groups / No	Identification	
Impacts/Endpoints	Health (physical, mental, developmental), environmental/ ecosystem, behavioral, economic, infrastructure, services, demographic, other	Assessment	
Health Endpoints	Identify health endpoints examined in HIA	Assessment	

*NOTE (August 2013): Changes were made to several Database Fields and Descriptions over the course of the HIA Review to refine the review framework. See Table 1 in the body of the report for an updated version of this table.

---

[1] Harris-Roxas and Harris (2011)
[2] Harris, Harris-Roxas, Harris, and Kemp (2007)

Database Field	Description/Examples	HIA Step	Addtl Info on Page(s)
Pathway of Impact	Air quality, community/household economics, education, exposure to hazards, healthcare access/insurance, housing, infectious disease, land use, lifestyle, mental health, mobility/access to services, noise pollution, nutrition, parks and recreation, physical activity, public health services, safety (personal, traffic, etc.) and security, social capital, soil quality, water quality, etc.	Assessment	16
Quantification of Impact	Direction (positive, negative, unclear, no effect), permanence, magnitude, likelihood (definite, probable, speculative, unlikely, uncertain), distribution/equity,[2,3]etc.	Assessment	17
Impact Prioritization	What methods/data were used to prioritize the impacts? **Note:** Prioritization of impacts often used to determine the impacts for which recommendations/mitigations will be developed Direction, permanence, magnitude, likelihood, distribution/equity of impacts may be used in prioritization (see Quantification [Characterization] of Impact)	Assessment	17
Decision-making Outcome	Describe the general outcome of the HIA, including recommendations, mitigations, etc.	Decision/Recommend	
HIA Report	(Attach HIA Report)	Decision/Recommend	18
Defensibility/Process Evaluation	Describe the quality of evidence and methodology; identify assumptions, limitations, barriers; etc.	Evaluation / Follow-up	
Effectiveness of HIA	Impact evaluation (direct, general, opportunistic, none [4]), health outcome evaluation (predictive accuracy, health impacts) / Undetermined **Note:** The effectiveness of the HIA cannot be determined by review of the HIA Report; this must be determined based on an internet/lit search	Evaluation / Follow-up	19
Follow-up Measures	Monitoring, health impact management, or other follow-up measures called for in the HIA / N/A	Evaluation / Follow-up	
Minimum Elements of HIA[5] Met? If no, what's missing	Yes / No - identify what's missing		20
GIS Used?	(If yes) Describe use – Illustrative, GIS analysis, etc. / No		
Environmental/Ecosystem Impacts Considered?	(If yes) Identify impacts / No		
Potential Improvements	Identify what could have potentially improved the HIA and/or its effectiveness. (Perhaps consult the HIA Practice Standards [5]) **Question to Consider:** How are the HIAs different and what could have been done to close the gap? For example, quantification of impacts (including costing); consideration of environmental/ecosystem impacts; additional information; use of GIS/spatial analysis; broader utilization of existing tools/models/resources (C-FERST/T-FERST, BenMAP, National Atlas of Ecosystem Services, EJ View, MyEnvironment, UCLA Health Impact Decision Support Tool, etc); consistency in conducting and reporting HIAs (e g , sector terminology, enhanced guidance/methodology, transparent/publicly-accessible documentation); clear reporting of recommendations and mitigations; identification of evaluation and follow-up measures; etc		21–24
Best Practices	Identify portions of the HIA process, report, etc. that stand out and describe these best practices. For example, tabular summary of potential impacts, including direction, extent, and populations most affected; defensibility of process; transparency of process documentation; etc * Potentially identify a set of HIAs within each sector representing the best of the best		

---

[3] Human Impact Partners (2011)
[4] Wismar, Blau, Ernst, and Figueras (2007)
[5] North American HIA Practice Standards Working Group (2010)

HEALTH IMPACT ASSESSMENT REVIEW
SECTORS

Observation: Sector terminology is not consistent between the various organizations promoting and/or reporting HIAs.

Resolution: The sectors to be analyzed will be defined using the terminology found in the Sustainable and Healthy Communities Research Program Draft Research Framework, June 1, 2011 (excerpts below).

---

* NOTE (August 2013): The HIA Review was conducted as part of the EPA's Sustainable and Healthy Communities Research Program and therefore utilized the sectors and sector descriptions identified for that research program. It should be noted that these sector descriptions do not necessarily reflect the sector descriptions typically utilized in the HIA field of practice.

TRANSPORTATION

Many communities find themselves in a transportation construct imposed by past generations' priorities, which doesn't easily fit more sustainable transportation models , imposes a high demand on fossil fuels and imposes an economic burdens on individuals and communities. In order to adequately transition to new transportation forms, decision makers must fully understand the full and long-term implications of new and transitional options. For example, many people think of mass transit alternatives only in terms of economic cost. However, there are many indirect economic and health benefits of mass transit that are relevant for decisions. It encourages better health by increasing walking, it raises adjacent property values, it lessens need for more destructive road building, decreases road congestion and so emissions, increases social capital and psychological health by enabling more incidental social interactions, and makes more jobs accessible to people who cannot afford cars. Also, adequate comparisons of transportation issues need to be placed in the proper context of alternatives, for example, two options for meeting commuting demand could be building a new highway lane or buying 50 buses for critical routes and creating incentives for ridership. The economic, environmental and social costs of these two alternatives are quite different, so tradeoffs should be clear.

Transportation issues also vary between rural and urban communities. Large, urban centers are usually growing, with increasing need for transportation capacity. However, they are also striving to decrease vehicles miles traveled (VMT) in cars by facilitating public transportation, walking and bicycling. Alternatively, rural residents are highly dependent on cars to access job centers and services. Most rural communities have a very limited public transportation infrastructure. While they recognize the value of public transportation from a sustainability perspective, it is less feasible for them because of cost, limited ridership, and the complexity of setting up regional partnerships with neighboring counties/cities. In addition, social and economic considerations affect transit use.

Illustrative Science Questions:
1. How can we assess a full accounting of environmental, economic and social effects of alternative transportation modes and fuels decisions on the sustainability and resilience of communities?
2. What new and existing community, state, and national policy options, incentives, interventions, or communication strategies can be used to improve transportation effects on community sustainability and resilience?
3. What suite of transportation options improves community sustainability and resilience most effectively and economically?

4. What associated land use and development designs can increase the use of public transportation systems?
5. How do transportation choices made in suburban and exurban areas affect overall community sustainability and resilience and the distribution of costs and benefits?
6. How can transportation design and choices affect a community's ability to adapt to climate change?
7. How can we communicate the full costs and benefits of transportation choices in a way that effectively informs decisions and changes behavior?

Health Impact Assessment (Pilot Review):
Pathways to a Healthy Decatur: A Rapid Health Assessment of the City of Decatur Community Transportation Plan
    Location: Decatur, Georgia
    Decision Making-Level: Local
    Organization(s): Georgia Tech Center for Quality Growth and Regional Development

HOUSING/BUILDINGS/INFRASTRUCTURE

Housing is a pivotal consideration for communities. Housing shortages (e.g., middle- or low-income, accommodations for families of local employers) have some communities focused on building more homes to attract new members to the community and to meet existing residents' needs. Other communities are driven by a desire to expand the tax base by renovating existing housing and commercial property. Still other communities suffer from abandoned buildings as their economy shrinks and people move away. Rising energy costs have made energy efficiency a top priority both for communities looking to cut costs in their subsidized housing programs and residents looking to make their money go farther, but "green" building is still a niche market and extensive energy retrofitting of existing homes can have a long payback. In addition, communities are making strides to revitalize their downtowns, but are challenged to address the cost and impact of sprawl caused by development.

Communities are increasingly developing land use plans and zoning decisions with sustainability goals in mind, but these can be undone easily by *ad hoc* rezoning or variances for specific buildings or developments the municipality thinks are economically justified or desired, based on the limited cost/benefit information available. Such decisions have long-term implications and can be made without the community participatory process that usually accompanies planning efforts. Roads will replace greenfield acreage, with the accompanying ecological services impacts. Other infrastructure will similarly displace vegetation, create impervious surfaces and impose long term costs. Buildings may be considered more expendable, and so not as permanent a decision, but seldom are buildings replaced by green space. So, unless a building is designed to be flexible for future use, unproductive buildings are often torn down, and embodied energy in buildings and materials are wasted.

With this kind of economic cost and environmental implications, community decisions that affect new development and infrastructure are best made with well-informed foresight, with sufficient information to allow accurate and timely comparison of long-term cumulative costs and benefits. In addition, there is significant new experience in retrofitting sprawl, dead malls and other underperforming sites into denser, mixed use, transit-oriented development, and full-cost accounting evaluation of these redevelopments would be useful. An accurate full cost/benefit evaluation comparing existing development, redevelopment and green field development implications would significantly benefit community decision making and could highlight significant economic opportunities.

Infrastructure evaluations also need to be made with full implications considered. There are many assumptions in infrastructure design that are being called to question by anecdotal evidence that can result in counterproductive decisions. For example, narrower streets in neighborhoods will better support walking, biking and a more protected feel, but wider streets are sometimes required to support larger fire trucks and allow faster response. An unforeseen consequence is that more firefighters are being killed in traffic accidents en route to fires than fighting fires. Such anecdotes highlight the need to evaluate the assumptions which underpin costly infrastructure decisions. An additional infrastructure issue raised by a community is that of transitioning from "grey" to "green" infrastructure. Making decisions on new infrastructure more easily incorporates comprehensive information than decisions on retrofitting green infrastructure or low impact development models into an existing built environment. But repair and retrofit decisions are regularly made in cities and would benefit from a comprehensive evaluation of gains and losses from different options.

Another example of the need for understanding the full implication of aging infrastructure issues is underground storage tanks for fuels. Ethanol can expand benzene plumes generated from leaking tanks, endangering groundwater supplies and causing vapor intrusion in buildings. A recent ORD analysis has shown that there is a significant population surrounding urban cores that have a higher vulnerability due to the co-location of gas stations and water supply wells. The SHCRP infrastructure, land use, transportation and site remediation research can significantly contribute to resolution of this issue.

Illustrative Science Questions:
1. What are the effects of the design, condition and maintenance practices of housing and other built environments (e.g. schools, office spaces, retail spaces, etc.) on human health and wellbeing? What are the best practices and products that communities can employ to minimize health risks (or promote wellness) from indoor exposures?
2. How can homes and infrastructure be designed and built to be more resilient to climate change and major environmental events that may be exacerbated by climate change (e.g. flooding, hurricanes, etc.)?
3. What type and mix of housing best promote the well-being of individuals and communities?
4. How does the distribution and type of built infrastructure affect the delivery of ecosystem services?
5. How can safe, affordable and healthful housing be distributed such that communities are better integrated and individuals have equitable access to the benefits associated with such housing (including community benefits such as access to transportation, education, healthy food, medical services and cultural amenities)?
6. How can building and infrastructure choices affect a community's ability to adapt to climate change?
7. What are the cumulative benefits of green practices implemented at the individual level (e.g. rain barrels, roof gardens, compact fluorescent bulbs, low VOC paint) in terms of improved health and well-being and increased delivery of ecosystem services? Which green practices contribute most to these benefits?

Health Impact Assessment (Pilot Review):
A Child Health Impact Assessment of the Massachusetts Rental Voucher Program
    Location: Massachusetts
    Decision Making-Level: State
    Organization(s): Boston University Child HIA Working Group

LAND USE

According to the participants in the community outreach efforts, many communities have developed "sustainability" plans, but these are often driven by more traditional long-standing planning practices that have mixed results. In addition, planning can take place without inclusive, well-informed discussions with community members and key stakeholders like local businesses and neighborhood associations. Rural communities are especially interested in ways to conduct planning that maintains their individual identities and land uses that sustain the rural economy (from agriculture to nature-based tourism on protected lands). Given the array of needs that must be considered (e.g., buildings, greenways, infrastructure), communities are challenged to know which sustainability practices and projects will serve them best. Communities are also interested in integrating land uses that promote healthy and safe lifestyles (e.g., greenways, trails, parks); however they are unclear about the benefits compared to impacts of their choices.

A holistic, cumulative assessment of all the costs and benefits of local decisions will allow fully-informed comparison of options and transparency of tradeoffs imposed by policy makers. For example, in attempts to increase their property tax base and create jobs, communities often permit big box stores on city fringes. However, a recent case study on tax values for different kinds of properties showed that infrastructure costs for big box stores would not be paid back by the expected revenue for decades, while urban mixed use midrise development payback was nearly 100 times greater per acre. Thus, a decision to permit a big box store may not have been made were the real costs and payback known. At the same time, there are unintended impacts and long term costs imposed by sprawl development that are not quantified, and so, not considered in such decisions, e.g. stormwater runoff pollution, heat island exacerbation, spreading of associated sprawl because of extended infrastructure, increased vehicle miles travelled (VMT) and mobile emissions, loss of green space or farmland and traffic problems.

Just as there are unaccounted costs in typical community decision making, there are also cumulative benefits for sustainable urbanism actions that are usually unrecognized or unquantified for decision making. For example, green space in a streetscape can treat stormwater, create walkable places for increased healthy lifestyles, feed biophilia and a feeling of well-being, increase social interactions and social capital, increase adjacent land values, support wildlife and pollinators, create activity-related jobs and increase customer traffic at adjacent businesses.

Besides the obvious parameters, planning and zoning decisions will affect traffic volume, viability of transit, feasibility of transportation alternatives like walkability and bikeability, proximity of services, proximity of green space, etc. These parameters of urban form affect health by affecting, for example, the amount of air pollution, the ability to incorporate exercise into daily living, safety (e.g. traffic accidents and crime) and the psychological benefits of increased social capital and freedom of mobility without a car. Communities are especially interested in the assessment of full costs and benefits of sprawl forms of growth compared to smart growth options. It is important that this be done with appropriate metrics for effective comparison. For example, GHG emissions per acre show a much different picture than GHG per capita, which demonstrates the energy efficiency of cities. Similarly, quantifying impervious surfaces in a dense, mixed use scenario may seem worse than in a diffuse suburban scenario, but comparing these scenarios on a common denominator of capacity basis (such as, "per 100,000 people" or "per 100 acres") will give a more realistic comparison of the ecosystem services impacts.

Community decision-making is often confounded by things that happen outside their boundaries and ability to control. Understanding what these processes are and how they factor into community problem solving is crucial for communities to move towards sustainable futures. For example, many communities in the Southeast share a common water supply and recent droughts have created "water wars" among communities. Communities also often export problems, e.g. the state of South Carolina's most impaired water body is located just downstream of Charlotte, NC.

Illustrative Science Questions:

1. What are the impacts of building density, mix (e.g., residential versus commercial/industrial), and location on the environmental, economic, and social health of a community?
2. What are the impacts of non-urban land use management (e.g., local versus distant agriculture, chemical use, crops and rotations, timber harvest), on the environmental, economic, and social health of a community?
3. How do a variety of land uses (e.g., community agriculture, parks, and urban services) contribute to community health and well-being and economic vitality?
4. What social and judicial levers or emerging information technology could compel behavior change related to land use at the individual and community levels?
5. How can we quantify the values of ecosystem services provided by a landscape and integrate those values with other social and economic parameters for improved decision making?
6. How do regional-scale processes (e.g. development outside community boundaries, air pollution transport, and shared water supplies) affect community-scale sustainability and how can these processes be factored into community decision-making?

Health Impact Assessment (Pilot Review):
Derby Redevelopment Health Impact Assessment
    Location: Historic Commerce City, Colorado
    Decision Making-Level: Local
    Organization(s): Tri-County Health Department

WASTE MANAGEMENT AND SITE REVITALIZATION

Communities may make decisions about waste management and disposal and site remediation and reuse options that are unsustainable because they lack the decision support tools needed to do comparative analysis of the short- and long-term costs and benefits in the life cycle of the various materials management options. Comprehensive analysis would incorporate elements such as, cumulative human exposures, carbon footprint, environmental justice, impairment of ecosystem services, transportation options and land use.

All communities are faced with managing a steady stream of municipal solid waste (MSW), the majority of which is managed in landfills. The per capita generation rates grew from 1960 to 1990 and remained in the range of 4.3 to 4.7 lb/person/day through 2009, when there was an apparent slight decline. Less than two-thirds of yard waste, aluminum cans and tires are recycled and less than a third of glass containers. Almost two-thirds of the entire MSW stream is organic materials that contain energy values, which are lost if the waste is disposed in a conventional landfill. Communities need to be able to evaluate the full costs and benefits of MSW recycling and management options in order to build the right infrastructure and set disposal fees and incentives in ways that meet costs and provide for future sustainable management. But even well-founded initiatives can fail if compliance by residents is easily circumvented. Bans on certain objects in landfills have been known to result in more of those objects thrown onto the roadside. Understanding motivations for non-compliance and incentives for compliance will assist communities to design and implement successful programs.

In addition to routine wastes, many communities periodically have to manage high volumes of debris from natural disasters, such as earthquakes, tornadoes, and flooding, often when infrastructure and communications are already strained. As an example, after a tornado ravaged Joplin, MO, the community needed support in managing the large volumes of debris that were generated. ORD was asked about comparative trade-offs in considering various risk management options, including landfilling and combustion. A decision support tool that identified the life cycle impacts of potential human health and ecosystem services of the various reuse/treatment/disposal options could have provided support in the recovery of the community. Additionally, these events (likely more frequent with climate change) create significant surges in materials that have the potential to create long-term environmental and/or human health impacts. Technological options and advance planning would allow communities to mobilize options that have the least impact on land resources, ecological resources, and human populations, particularly populations that are more susceptible or already disproportionally exposed to environmental stressors. Other man-made debris, such as building and demolition discards and roadbuilding and maintenance wastes, could also be recycled and disposed of more sustainably with the right technologies, planning, ordinances, and pricing schedules.

Communities have more blighted properties following the recent economic recession. Some of these, particularly defunct commercial and industrial properties, are suspected of being contaminated by hazardous constituents. Brownfields grants and other voluntary programs can help communities assess and redevelop properties. Added focus on the full costs and benefits of redevelopment alternatives (park, green space, urban agriculture, business park, small business, residential, green energy) could maximize sustainability gains. Redevelopment also offers opportunities to rectify conditions faced by disadvantaged populations.

This research will be integrated with the Office of Solid Waste and Emergency Response's (OSWER) advice to individuals and communities for managing MSW (http://www.epa.gov/wastes/nonhaz/municipal/) and activities on Brownfields and land revitalization. Note that this objective is complemented by research in Objectives 3a and 3b. Whereas Objective 1e research focuses on community decision-making, the focus of the Theme 3 objectives is more on the R&D to ensure that environmental regulations involving waste and materials management and site remediation are based on sound science and engineering.

Illustrative Science Questions:
1. What are the most important decisions facing communities in the area of waste disposal, materials management, and site remediation?
2. How can we improve the decision process to assist communities in managing debris after extreme weather events?
3. What are the direct, indirect, and cumulative effects of the most commonly faced decisions about waste disposal, materials management, and site remediation options on human health (including children and the elderly) and ecosystem services?
4. What are the likely economic consequences of these options, including economic multipliers, lost resources, and changes in property valuation, economic stability and job creation?
5. What are the likely social consequences of these effects, including social acceptance of possible actions, effect on social capital and environmental justice of actions and outcomes?

Health Impact Assessment (Pilot Review):
Health Impact Assessment of NRMT's Request for a Special Use Permit
    Location: Bernalillo County, New Mexico
    Decision Making-Level: County
    Organization(s): Bernalillo County Place Matters Team, New Mexico Health Equity Working Group

HEALTH IMPACT ASSESSMENT REVIEW
HIA TYPE

Forms of health impact assessment

	Mandated	Decision-support	Advocacy	Community-led
Description	Occurs in the context of an environmental impact assessment (EIA), integrated impact assessment (IIA) or environmental, social and health impact assessment (ESHIA) and is done to meet a regulatory or statutory requirement	Conducted voluntarily by, or with the agreement of, organisations responsible for a proposal, with the goal of improving decision-making and implementation	Conducted by organisations or groups who are neither proponents or decision-makers, with goal of influencing decision-making and implementation	Conducted by potentially affected communities on issues or proposals that are of concern
Purpose	• Meeting a regulatory or statutory requirement • Minimising negative health impacts	• Improving decision-making and implementation • Minimising negative health impacts • Maximising positive health impacts	• Ensuring under-recognised health concerns are addressed in design, decision-making and implementation • Minimising negative health impacts • Maximising positive health impacts	• Ensuring the community's health-related concerns are identified and addressed • Enabling greater participation of communities in decisions that affect them • Minimising negative health impacts • Maximising positive health impacts
Origins	Environmental health	Environmental health, social view of health, health equity	Social view of health, health equity	Social view of health, health equity
Role of values and judgements	Almost no role for values in assessment, judgements often not acknowledged	Implied role for values and judgements	More explicit role for values and judgements	Driven by community values and judgements
Conducted by	Consultants	Government agencies, consultants	Non-governmental organisations (NGOs), universities, other agencies	Communities, often aided by HIA practitioners in NGOs, universities or other agencies
Resourced by	Proponents	Government agencies	Varied	Communities themselves
Overseen by	Proponents	Government agencies	Varied	Communities themselves
Role of stakeholders	Providing technical information	Informing the assessment	Guiding the assessment	Controlling and conducting the assessment
Type of learning	Technical	Technical/conceptual	Conceptual/social	Social

Taken from Harris-Roxas and Harris. 2011. Differing forms, differing purposes: A typology of health impact assessment

HEALTH IMPACT ASSESSMENT REVIEW
HIA RIGOR

DESK BASED	RAPID	INTERMEDIATE	COMPREHENSIVE
2-6 weeks for one person full time[1].	6 to 12 weeks for one person full time.	12 weeks to 6 months for one person full time.	6 to 12 months for one person full time.
Provides a broad overview of potential health impacts.	Provides a more detailed overview of potential health impacts.	Provides a more thorough assessment of potential health impacts, and more detail on specific predicted impacts.	Provides a comprehensive assessment of potential health impacts.
Could be used where time and resources are limited.	Could be used where time and resources are limited.	Requires significant time and resources.	Requires significant time and resources.
Is an 'off the shelf' exercise based on collecting and analysing existing accessible data.	Involves collecting and analysing existing data with limited input from experts and key stakeholders	Involves collecting and analysing existing data as well as gathering new qualitative data from stakeholders and key informants.	Involves collecting and analysing data from multiple sources (qualitative and quantitative)
Activities include accessing off the shelf resources and synthesising and appraising information.	Activities include accessing resources, hosting and supporting meetings, and synthesising and appraising information. If capacity does not exist in-house, consideration should be given to commissioning external assessors.	Activities include accessing resources, hosting and supporting meetings, identifying stakeholders and key informants, gathering and analysing qualitative and quantitative data, and synthesising and appraising information. If capacity does not exist in-house, consideration should be given to commissioning external assessors.	Activities include accessing resources, hosting and supporting meetings, identifying stakeholders and key informants, gathering and analysing qualitative and quantitative data, and synthesising and appraising information. If capacity does not exist in-house, consideration should be given to commissioning external assessors.

LESS IMPACTS ━━━━━━━━━━━━━━━━━━━━━━━━━▶ **MORE IMPACTS**

[1] The time involved will vary depending on the number of people actively involved in undertaking HIA tasks. For example a comprehensive assessment may take a team of four people three months to complete.

Level of HIA and number and depth of impacts to assess:			
DESK BASED	**RAPID**	**INTERMEDIATE**	**COMPREHENSIVE**
No more than three impacts, assessed in less detail	No more than three impacts, assessed in more detail	Three to ten impacts, assessed in detail	All potential impacts, assessed in detail
Provides a broad overview of potential health impacts	Provides a more detailed overview of potential health impacts	Provides a more thorough assessment of potential health impacts, and more detail on specific predicted impacts	Provides a comprehensive assessment of potential health impacts

Taken from Harris, Harris-Roxas, Harris, and Kemp. 2007. Health Impact Assessment: A Practical Guide.

HEALTH IMPACT ASSESSMENT REVIEW
DATA TYPES/MAJOR DATA SOURCES

DATA TYPES

Data types are general categories of information – models, literature, websites, and data.

MAJOR DATA SOURCES

It is not necessary to list all data sources, journals, authors, etc. referenced in the HIA, but the major sources of data utilized in analysis, must be identified by name and where applicable, type of data.

If identified, enter the year, scale, and type of data; see below for example:

> HUD American Housing Survey, HUD 2005 Fair Market Rent, HUD Special Tabulations of Households (by income, tenure, age of householder, and housing conditions; 2005), City of Boston Homeless Census (2004)

A common data source used in HIA comes from the U.S. Census Bureau. See the types of census data and the various scales at which the data are reported below:

Types of Census Data – Age, agriculture, births, business establishments, communications, construction, cost of living, crime, deaths, education, elections, employment, energy, finance, government, health, households, housing, immigration, income, manufactures, marriages and divorces, media, natural resources, population, poverty, race and Hispanic origin, residence, retail sales, science and engineering, social services, tourism, transportation, veterans, etc.

Scale of Census Data – National, regional, state, metropolitan area, county (or equivalent), city/township, school district, census block, etc.

HEALTH IMPACT ASSESSMENT REVIEW
PATHWAY OF IMPACT

PATHWAY OF IMPACT

The pathway of impact is the pathway through which the proposed policy, program, or project is expected to affect health.

COMMON PATHWAYS

Below is a list of common pathways used in HIAs. To the extent possible, use this terminology when identifying the pathways of impact; additional description can be added in parentheses for more general pathways. If a pathway of impact was examined in an HIA and is not included in this list, be sure to include it in the data entry, being consistent in terminology across HIAs (if applicable).

- Air quality
- Community/household economics
- Education
- Exposure to hazards (pollutants, health hazards, etc.)
- Healthcare access/insurance
- Housing (physical housing conditions, affordability, housing instability, etc.)
- Infectious disease
- Land use
- Lifestyle
- Mental health
- Mobility/access to services
- Noise pollution
- Nutrition
- Parks and recreation/green space
- Physical activity
- Public health services
- Safety and security
- Social capital
- Soil quality
- Traffic safety
- Water quality

Descriptions of many of these pathways and their common downstream health endpoints/effects are available at: http://www.hiaguide.org/sectors-and-causal-pathways/pathways

HEALTH IMPACT ASSESSMENT REVIEW
QUANTIFICATION OF IMPACT/IMPACT PRIORITIZATION

> * NOTE (August 2013): In the final master database, the Quantification of Impact field will be revised to Characterization of Impact to more accurately reflect the actual process of impact assessment, which can involve both qualitative and quantitative characterization.

DIRECTION

Positive = Changes that may improve health
Negative = Changes that may detract from health
Unclear = Unknown how health will be impacted
No effect = No effect on health

PERMANENCE (*severity*)

Low = Causes impacts that can be quickly and easily managed or do not require treatment
Medium = Causes impacts that necessitate treatment or medical management and are reversible
High = Causes impacts that are chronic, irreversible or fatal

MAGNITUDE (*relative to population size*)

Low = Causes impacts to no or very few people
Medium = Causes impacts to wider number of people
High = Causes impacts to many people

LIKELIHOOD

Definite = impacts will occur as a result of the proposal
Probable = it is likely that impacts will occur as a result of the proposal
Speculative = it is possible that impacts will occur as a result of the proposal
Unlikely = it is unlikely that impacts will occur as a result of the proposal
Uncertain = it is unclear if impacts will occur as a result of the proposal

DISTRIBUTION/EQUITY OF IMPACT

Name subpopulation impacted more (e.g., "low-income residents impacted more"; "Blacks impacted more") or "equal impacts"

Taken/adapted from Human Impact Partners. 2011. A Health Impact Assessment Toolkit: A handbook to conducting HIA (3rd edition).

HEALTH IMPACT ASSESSMENT REVIEW
HIA REPORT

DATA ENTRY – HIA REPORT

Attach the reviewed HIA Report to the database record using the following steps:

1. Double-click in the HIA Report field of the Data Entry Form.

2. In the Attachments dialog box that pops up, click on the "Add" button at the top right.

3. Navigate to the HIA Reports located on the network:

   L:\Public\NERL-PUB\Health Impact Assessment\HIA Review Materials\HIA Reports

4. Click on the appropriate HIA Report so that the file name is highlighted, and then click on the "Open" button at the bottom right of the window.

   Note: The HIA Report file should now show up in the Attachments dialog box.

5. Click the "OK" button at the bottom of the dialog box and the file should display in the HIA Report field of the Data Entry Form.

HEALTH IMPACT ASSESSMENT REVIEW
EFFECTIVENESS OF HIA

IMPACT EVALUATION

Direct effectiveness –   a decision is dropped or modified as a result of the HIA.

General effectiveness –   the assessment was considered adequately by the decision-makers, but does not result in modifications to the proposed decision.

Opportunistic effectiveness – the HIA is conducted because it is assumed that it will support the proposed decision.

None (ineffectiveness) –   decision-makers do not take account of the assessment

| | | Modification of pending decisions according to health/equity/community aspects and inputs | |
		Yes	No
Health/equity/ community adequately acknowledged	Yes	Direct effectiveness • HIA-related changes in the decision • due to the HIA the project was dropped • decision was postponed	General effectiveness • reasons provided for not following HIA recommendations • health consequences are negligible or positive • HIA has raised awareness among policy-makers
	No	Opportunistic effectiveness • the decision would have been made anyway	No effectiveness • the HIA was ignored • the HIA was dismissed

Taken from Wismar, Blau, Ernst, and Figueras. 2007. The Effectiveness of Health Impact Assessment: Scope and Limitations of Supporting Decision-making in Europe.

HEALTH IMPACT ASSESSMENT REVIEW
MINIMUM ELEMENTS OF HIA

MINIMUM ELEMENTS

A health impact assessment (HIA) must include the following minimum elements, which together distinguish HIA from other processes. An HIA:

1. Is initiated to inform a decision-making process, and conducted in advance of a policy, plan, program, or project decision;

2. Utilizes a systematic analytic process with the following characteristics:

    2.1 Includes a scoping phase that comprehensively considers potential impacts on health outcomes as well as on social, environmental, and economic health determinants, and selects potentially significant issues for impact analysis;
    2.2 Solicits and utilizes input from stakeholders;

    2.3 Establishes baseline conditions for health, describing health outcomes, health determinants, affected populations, and vulnerable sub-populations;

    2.4 Uses the best available evidence to judge the magnitude, likelihood, distribution, and permanence of potential impacts on human health or health determinants;

    2.5 Rests conclusions and recommendations on a transparent and context-specific synthesis of evidence, acknowledging sources of data, methodological assumptions, strengths and limitations of evidence and uncertainties;

3. Identifies appropriate recommendations, mitigations and/or design alternatives to protect and promote health;
4. Proposes a monitoring plan for tracking the decision's implementation on health impacts/determinants of concern;

5. Includes transparent, publicly-accessible documentation of the process, methods, findings, sponsors, funding sources, participants and their respective roles.

Taken from North American HIA Practice Standards Working Group. 2010. Minimum Elements and Practice Standards for Health Impact Assessment.

HEALTH IMPACT ASSESSMENT REVIEW
POTENTIAL IMPROVEMENTS/BEST PRACTICES

To identify areas of potential improvements and even best practices, it may be helpful to consult the HIA Practice Standards.

Taken from North American HIA Practice Standards Working Group. 2010. Minimum Elements and Practice Standards for Health Impact Assessment

HIA PRACTICE STANDARDS

Adherence to the following standards is recommended to advance effective HIA practice:

1. **General standards for the HIA process**

    1.1 An HIA should include, at a minimum, the stages of **screening, scoping, assessment, recommendations,** and **reporting** described below.

    1.2 **Monitoring** is an important follow-up activity in the HIA process. The HIA should include a follow-up monitoring plan to track the outcomes of a decision and its implementation.

    1.3 **Evaluation** of the HIA process and impacts is necessary for field development and practice improvement. Each HIA process should begin with explicit, written goals that can be evaluated as to their success at the end of the process.

    1.4 HIA should respect the needs and timing of the decision-making process it evaluates.

    1.5 HIA requires integration of knowledge from many disciplines; the practitioner or practitioner team must take reasonable and available steps to identify, solicit and utilize the expertise, including from the community, needed to both identify and answer questions about potentially significant health impacts.

    1.6 Meaningful and inclusive stakeholder participation (e.g., community, public agency, decision-maker) in each stage of the HIA supports HIA quality and effectiveness. Each HIA should have a specific engagement and participation approach that utilizes available participatory or deliberative methods suitable to the needs of stakeholders and context.

    1.7 HIA is a forward looking activity intended to inform an anticipated decision; however, HIA may appropriately conduct or utilize analysis, or evaluate an existing policy, project or plan to prospectively inform a contemporary decision or discussion.

    1.8 Where integrated impact assessment is required and conducted, and requirements for impact assessment include responsibility to analyze health impacts, HIA should be part of an integrated impact assessment process to advance efficiency, to allow for interdisciplinary analysis and to maximize the potential for advancing health promoting mitigations or improvements.

    1.9 HIA integrated within another impact assessment process should adhere to these practice standards to the greatest extent possible.

2. **Standards for the screening stage**

   2.1 Screening should clearly identify all the decision alternatives under consideration by decision-makers at the time the HIA is considered.

   2.2 Screening should determine whether an HIA would add value to the decision-making process. The following factors may be among those weighed in the screening process:

   > 2.2.1 The potential for the decision to result in substantial effects on public health, particularly those effects which are avoidable, involuntary, adverse, irreversible or catastrophic

   > 2.2.2 The potential for unequally distributed impacts

   > 2.2.3 Stakeholder and decision-maker concerns about a decision's health effects

   > 2.2.4 The potential for the HIA to result in timely changes to a policy plan, policy or program

   > 2.2.5 The availability of data, methods, resources and technical capacity to conduct analyses

   > 2.2.6 The availability, application, and effectiveness of alternative opportunities or approaches to evaluate and communicate the decision's potential health impacts

   2.3 Sponsors of the HIA should document the explicit goals of the HIA and should notify, to the extent feasible, decision-makers, identified stakeholders, affected individuals and organizations, and responsible public agencies on their decision to conduct an HIA.

3. **Standards for the scoping phase**

   3.1 Scoping of health issues and public concerns related to the decision should include identification of: 1) the decision and decision alternatives that will be studied; 2) potential significant health impacts and their pathways (e.g., a logic model); 3) research questions for impact analysis; 4) demographic, geographical and temporal boundaries for impact analysis; 5) evidence sources and research methods expected for each research question in impacts analysis; 6) the identity of vulnerable subgroups of the affected population; 7) an approach to the evaluation of the distribution of impacts; 8) roles for experts and key informants; 9) the standards or process, if any, that will be used for determining the significance of health impacts; 10) a plan for external and public review; and 11) a plan for dissemination of findings and recommendations.

   3.2 The scoping process should establish the individual or team responsible for conducting the HIA and should define their roles.

   3.3 Scoping should include consideration of all potential pathways that could reasonably link the decision and/or proposed activity to health, whether direct, indirect, or cumulative.

   3.4 The consideration of potential pathways should be informed by the expertise and experience of assessors as well as perspectives of the affected communities, health officials and decision-makers. The assessment team should solicit input from public health officials and local medical practitioners to ensure adequate representation by the entities responsible for and knowledgeable about health conditions. The assessment team should solicit input from members of affected communities or representative organizations via public meetings, written comments, or interviews to understand their views and concerns. The assessment team should solicit input from decision-makers to understand their views on the decision's relationship to health.

3.5 The final scope should focus on those impacts with the greatest potential significance, with regards to factors including but not limited to magnitude, certainty, permanence, stakeholder priorities, and equity.

3.6 The scope should include an approach to evaluate any potential inequities in impacts based on population characteristics, including but not limited to age, gender, income, place (disadvantaged locations), and race or ethnicity.

3.7 The HIA scoping process should identify a mechanism to incorporate new, relevant information and evidence into the scope as it becomes available, including through expert or stakeholder feedback.

4. **Standards for the assessment phase**

4.1 Assessment should include, at a minimum, a baseline conditions analysis and qualified judgments of potential health impacts:

4.1.1 Documentation of baseline conditions should include the documentation of both population health vulnerabilities (based on the population characteristics described above) and inequalities in health outcomes among subpopulations or places.

4.1.2 Evaluation of potential health impacts should be based on a synthesis of the best available evidence, as qualified below.

4.1.3 To support determinations of impact significance, the HIA should characterize health impacts according to characteristics such as direction, magnitude, likelihood, distribution within the population, and permanence.

4.2 Judgments of health impacts should be based on a synthesis of the best available evidence. This means:

4.2.1 Evidence considered may include existing data, empirical research, professional expertise and local knowledge, and the products of original investigations.

4.2.2 When available, practitioners should utilize evidence from well-designed and peer-reviewed systematic reviews.

4.2.3 HIA practitioners should consider published evidence, both supporting and refuting particular health impacts.

4.2.4 The expertise and experience of affected members of the public (local knowledge), whether obtained via the use of participatory methods, collected via formal qualitative research methods, or reflected in public testimony, is potential evidence.

4.2.5   Justification for the selection or exclusion of particular methodologies and data sources should be made explicit (e.g., resource constraints).

4.2.6   The HIA should acknowledge when available methods were not utilized and why (e.g., resource constraints).

4.3 Impact analysis should explicitly acknowledge methodological assumptions as well as the strengths and limitations of all data and methods used.

4.3.1   The HIA should identify data gaps that prevent an adequate or complete assessment of potential impacts.

4.3.2   Assessors should describe the uncertainty in predictions.

4.3.3   Assumptions or inferences made in the context of modeling or predictions should be made explicit.

4.4 The lack of formal, scientific, quantitative or published evidence should not preclude reasoned predictions of health impacts.

5. **Standards for the recommendations phase**

5.1 The HIA should include specific recommendations to manage the health impacts identified, including alternatives to the decision, modifications to the proposed policy, program, or project, or mitigation measures.

5.2 Where needed, expert guidance should be utilized to ensure recommendations reflect current effective practices.

5.3 The following criteria may be considered in developing recommendations and mitigation measures: responsiveness to predicted impacts; specificity; technical feasibility; enforceability; and authority of decision-makers.

5.4 Recommendations may include those for monitoring, reassessment, and adaptations to help manage uncertainty in impact assessment.

6. **Standards for the reporting phase**

6.1 The responsible parties should complete a report of the HIA findings and recommendations.

6.2 To support effective, inclusive communication of the principal HIA findings and recommendations, a succinct summary should be created that communicates findings in a way that allows all stakeholders to understand, evaluate, and respond to the findings.

6.3 The full HIA report should document the screening and scoping processes and identify the sponsor of the HIA and the funding source, the team conducting the HIA, and all other participants in the HIA and their roles and contributions. Any potential conflicts of interest should be acknowledged.

6.4 The full HIA report should, for each specific health issue analyzed, discuss the available scientific evidence, describe the data sources and analytic methods used for the HIA including their rationale, profile existing conditions, detail the analytic results, characterize the health impacts and their significance, list corresponding recommendations for policy, program, or project alternatives, design or mitigations, and describe the limitations of the HIA.

6.5 Recommendations for decision alternatives, policy recommendations, or mitigations should be specific and justified. The criteria used for prioritization of recommendations should be explicitly stated and based on scientific evidence and, ideally, informed by an inclusive process that accounts for stakeholder values.

6.6 Distribute HIA and/or findings to stakeholders that were involved in the HIA. The HIA reporting process should offer stakeholders and decision-makers a meaningful opportunity to critically review evidence, methods, findings, conclusions, and recommendations. Ideally, a draft report should be made available and readily accessible for public review and comment. The HIA practitioners should address substantive criticisms either through a formal written response or HIA report revisions before finalizing the HIA report.

6.7 The final HIA report should be made publicly accessible.

7. **Standards for the monitoring phase**

7.1 The HIA should include a follow-up monitoring plan to track the decision outcomes as well as the effect of the decision on health impacts and/or determinants of concern.

7.2 The monitoring plan should include: 1) goals for short- and long-term monitoring; 2) outcomes and indicators for monitoring; 3) lead individuals or organizations to conduct monitoring; 4) a mechanism to report monitoring outcomes to decision-makers and HIA stakeholders; 5) triggers or thresholds that may lead to review and adaptation in decision implementation; and 6) identified resources to conduct, complete, and report the monitoring.

7.3 Where possible, recommended mitigations should be further developed and integrated into an HIA (or other) management plan, which clearly outlines how each mitigation measure will be implemented. Management plans commonly include information on: deadlines, responsibilities, management structure, potential partnerships, engagement activities and monitoring and evaluation related to the implementation of the HIA mitigations. For greater effectiveness, HIA management plans should be developed in collaboration with, or at least with the input from, the entity responsible for implementing the plan. Management plans are living documents that will need to be revised and improved on an on-going basis.

7.4 When monitoring is conducted, methods and results from monitoring should be made available to the public.

Appendix B – Master List of HIAs Reviewed

ID	HIA Title	Sector
1	Pathways to a Healthy Decatur: A Rapid Health Assessment of the City of Decatur Community Transportation Plan	Transportation
2	Affordable Housing and Child Health: A Child Health Impact Assessment of the Massachusetts Rental Voucher Program	Housing/Buildings/ Infrastructure
3	Health Impact Assessment - Derby Redevelopment, Historic Commerce City, Colorado	Land Use
4	Health Impact Assessment of NRMT's Request for a Special Use Permit	Waste Management/ Site Revitalization
5	The Red Line Transit Project Health Impact Assessment	Transportation
6	Health Impact Assessment Report: Alcohol Environment - Village of Weston, WI	Land Use
7	Eastern Neighborhoods Community Health Impact Assessment Final Report	Land Use
8	Health Impact Assessment: South Lincoln Homes, Denver CO	Housing/Buildings/ Infrastructure
9	Spokane University District Pedestrian/Bicycle Bridge Health Impact Assessment	Transportation
10	Health Impact Assessment: An Analysis of Potential Sites for a Regional Recreation Center to Serve North Aurora, Colorado	Land Use
11	The Impact of U.S. Highway 550 Design on Health and Safety in Cuba, New Mexico: A Health Impact Assessment	Transportation
12	Community Health Assessment: Bernal Heights Preschool - An Application of the Healthy Development Measurement Tool (HDMT)	Housing/Buildings/ Infrastructure
13	St. Louis Park Comprehensive Plan - Health Impact Assessment	Land Use
14	Comprehensive Health Impact Assessment: Clark County Bicycle and Pedestrian Master Plan	Transportation
15	Health Impact Assessment: Key Recommendations of the Northeast Area Plan	Land Use
16	Yellowstone County/City of Billings Growth Policy Health Impact Assessment	Land Use
17	Health Impact Assessment, June 20, 2011: Duluth, Minnesota's Complete Streets Resolution, Mobility in the Hillside Neighborhoods and the Schematic Redesign of Sixth Avenue East	Transportation
18	Knox County Health Department Community Garden Health Impact Assessment: Recommendations for Lonsdale, Inskip and Mascot	Land Use
19	Alaska Outer Continental Shelf - Beaufort Sea and Chukchi Sea Planning Areas, Oil and Gas Lease Sales 209, 212, 217, and 221 Draft Environmental Impact Statement; Appendix J - Public Health	Land Use
20	Divine Mercy Development Health Impact Assessment	Land Use
21	Fort McPherson Rapid Health Impact Assessment: Zoning for Health Benefit to Surrounding Communities During Interim Use	Land Use
22	Re: November 10th Merced County General Plan Update (MCGPU) Preferred Growth Alternative Decision	Land use
23	Health Impact Assessment of Modifications to the Trenton Farmer's Market (Trenton, New Jersey)	Housing/Buildings/ Infrastructure
24	SE 122nd Avenue Planning Study Health Impact Assessment	Land Use
25	Concord Naval Weapons Station Reuse Project Health Impact Assessment	Waste Management/ Site Revitalization

ID	HIA Title	Sector
26	Health Impact Assessment: Hawai'i County Agriculture Development Plan	Land Use
27	Mass Transit Health Impact Assessment: Potential Health Impacts of the Governor's Proposed Redirection of California State Transportation Spillover Funds	Transportation
28	The Rental Assistance Demonstration Project - A Health Impact Assessment	Housing/Buildings/ Infrastructure
29	Case Study: Bloomington Xcel Energy Corridor Trail Health Impact Assessment	Land Use
30	Jack London Gateway Rapid Health Impact Assessment: A Case Study	Housing/Buildings/ Infrastructure
31	Health Impact Assessment for Proposed Coal Mine at Wishbone Hill, Matanuska-Susitna Borough Alaska (DRAFT)	Land Use
32	HIA of the Still/Lyell Freeway Channel in the Excelsior District	Transportation
33	Neenah-Menasha Sewerage Commission Biosolids Storage Facility, Greenville, WI	Waste Management/ Site Revitalization
34	City of Ramsey Health Impact Assessment	Land Use
35	A Health Impact Assessment of Accessory Dwelling Unit Policies in Rural Benton County, Oregon	Housing/Buildings/ Infrastructure
36	The Health Impact Assessment (HIA) of the Commonwealth Edison (ComEd) Advanced Metering Infrastructure (AMI) Deployment	Housing/Buildings/ Infrastructure
37	Atlanta Beltline Health Impact Assessment	Land Use
38	Zoning for a Healthy Baltimore: A Health Impact Assessment of the Transform Baltimore Zoning Code Rewrite	Land Use
39	Hood River County Health Department Health Impact Assessment for the Barrett Property	Land Use
40	Unhealthy Consequences: Energy Costs and Child Health - A Child Health Impact Assessment of Energy Costs and the Low Income Home Energy Assistance Program	Housing/Buildings/ Infrastructure
41	Technical Report 9: Highway 99 Sub-Area Plan Health Impact Assessment	Land Use
42	Columbia River Crossing Health Impact Assessment	Transportation
43	Inupiat Health and Proposed Alaskan Oil Development: Results of the First Integrated Health Impact Assessment/Environmental Impact Statement for Proposed Oil Development on Alaska's North Slope	Land Use
44	Page Avenue Health Impact Assessment	Land Use
45	Pittsburg Railroad Avenue Specific Plan Health Impact Assessment	Land Use
46	The Sellwood Bridge Project: A Health Impact Assessment	Transportation
47	The East Bay Greenway Health Impact Assessment	Land Use
48	A Rapid Health Impact Assessment of the City of Los Angeles' Proposed University of Southern California Specific Plan	Housing/Buildings/ Infrastructure
49	Taylor Energy Center Health Impact Assessment	Land Use
50	Anticipated Effects of Residential Displacement on Health: Results from Qualitative Research	Housing/Buildings/ Infrastructure
51	Rapid Health Impact Assessment, Crook County/City of Prineville, Bicycle and Pedestrian Safety Plan	Transportation
52	29th St. / San Pedro St. Area Health Impact Assessment	Housing/Buildings/ Infrastructure
53	SR 520 Health Impact Assessment: A Bridge to a Healthier Community	Transportation
54	A Health Impact Assessment on Policies Reducing Vehicle Miles Traveled in Oregon Metropolitan Areas	Transportation

ID	HIA Title	Sector
55	Health Effects of Road Pricing In San Francisco, California: Findings from a Health Impact Assessment	Transportation
56	Santa Monica Airport Health Impact Assessment	Transportation
57	Lowry Corridor, Phase 2 Health Impact Assessment	Housing/Buildings/ Infrastructure
58	Battlement Mesa Health Impact Assessment (2nd Draft)	Land Use
59	Douglas County Comprehensive Plan Update Health Impact Assessment	Land Use
60	The Executive Park Subarea Plan Health Impact Assessment: An Application of the Healthy Development Measurement Tool (HDMT)	Land Use
61	Hospitals and Community Health HIA: A Study of Localized Health Impacts of Hospitals	Housing/Buildings/ Infrastructure
62	Health Impact Assessment on Transportation Policies in the Eugene Climate and Energy Action Plan	Transportation
63	Oak to Ninth Avenue Health Impact Assessment	Land Use
64	A Health Assessment of Mixed Use Redevelopment Nodes and Corridors in Lincoln, Nebraska	Land Use
65	Health Impact Assessment (HIA) of Proposed "Road Diet" and Restriping Project on Daniel Morgan Avenue in Spartanburg, South Carolina	Transportation
66	Treasure Island Community Transportation Plan	Transportation
67	Healthy Tumalo Community Plan: A Health Impact Assessment on the Tumalo Community Plan; A Chapter Of The 20-Year Deschutes County Comprehensive Plan Update	Land Use
68	Strategic Health Impact Assessment on Wind Energy Development in Oregon (Public Review Draft)	Land Use
69	Impacts on Community Health of Area Plans for the Mission, East SoMa, and Potrero Hill/Showplace Square: An Application of the Healthy Development Measurement Tool	Land Use
70	Pathways to Community Health: Evaluating the Healthfulness of Affordable Housing Opportunity Sites Along the San Pablo Avenue Corridor Using Health Impact Assessment	Housing/Buildings/ Infrastructure
71	MacArthur BART Transit Village Health Impact Assessment	Land Use
72	Healthy Corridor for All: A Community Health Impact Assessment of Transit-oriented Development Policy in St. Paul Minnesota	Land Use
73	Health Impact Assessment Point Thomson Project	Land Use
74	Assessment of Open Burning Enforcement in La Crosse County	Waste Management/ Site Revitalization
75	Interstate 75 Focus Area Study Health Impact Assessment	Transportation
76	A Rapid Health Impact Assessment of the Long Beach Downtown Plan	Housing/Buildings/ Infrastructure
77	Humboldt County General Plan Update Health Impact Assessment	Land Use
78	Rapid Health Impact Assessment: Vancouver Comprehensive Growth Management Plan 2011	Land Use
79	Lake Oswego to Portland Transit Project: Health Impact Assessment	Transportation
80	HOPE VI to HOPE SF San Francisco Public Housing Redevelopment: A Health Impact Assessment	Housing/Buildings/ Infrastructure
81	Health Impact Assessment of the Port of Oakland	Transportation

Appendix C – Quality Assurance Review Documentation

Health Impact Assessment QA Review
Follow-up/Preliminary Corrective Action Meeting Notes
July 3, 2012

Summary

There were considerable discrepancies in data entry for the health impact assessments (HIAs) designated for quality assurance (QA) review. This may be due, in part, to the unfamiliarity of the reviewers with HIAs in general, inexperience in applying the principles of the review process, as well as the timing of the QA Review implementation in the overall review, as the HIAs in the QA Review were the first three HIAs reviewed by the reviewers.

Data entry discrepancies between the initial review and the QA Review of each HIA were identified and the discrepancies collectively reviewed to identify overall trends. The areas for improvement (shown below); challenges, questions, and lessons learned from the QA Review; and the review pathforward will be discussed.

I. Areas for Improvement

1) Use specified terminology (HIA Review Guidelines, pages 6–7; *highlighted in gray*)

Decision-making Level	HIA Rigor	Quantification of Impact
Organization Type	Source of Evidence	Effectiveness of HIA*
Contact	Data Types	GIS Used?
Sector(s)	Impacts/Endpoints	
HIA Type	Pathway of Impact	

> * NOTE (August 2013): In the final master database, the Quantification of Impact field will be revised to Characterization of Impact to more accurately reflect the actual process of impact assessment; impacts in HIAs can be judged qualitatively or quantitatively.

2) Consistency in use of specified terminology

Sector(s) – see SHCRP definitions in HIA Review Guidelines, pages 8–12

HIA Type – see HIA Review Guidelines, page 13
- Advocacy vs. Decision Support – "Outagamie County Public Health first learned of the proposed biosolids storage facility and the *concerns of the community*… Outagamie County Public Health Division *has no regulatory authority* for biosolids production, transport, storage or use. Outagamie County Public Health Division's *sole interest in this project is to review potential health concerns and propose methods to reduce those risks*…It was also concluded that the HIA would provide a background that could serve as a base to address citizen inquiries and complaints that local agencies may encounter if the storage facility is built." – Advocacy

HIA Rigor – see HIA Review Guidelines, page 14; three main factors – number of impacts assessed, level/depth of assessment, and length of time (but time involved will vary depending on the number of people and effort actually involved; times noted are for one full-time person)
- Desk-based vs. Rapid – "The impact assessment was completed and presented to the community in less than one month. Outagamie County Public Health staff *attended both meetings* to learn more about the health concerns being raised. In order to narrow the focus on the most significant potential health impacts… the

following activities were completed: *interviews with existing biosolids storage facilities in Outagamie County...*" – Rapid

- Intermediate vs. Comprehensive – "This assessment does not use all the indicators from each of the six HDMT elements, but rather focuses on four elements that captured the possible impacts of ADUs in Benton County. The elements used are healthy housing, access to goods and services, social and family cohesion, and transportation and mobility. Accessory dwelling units *likely have impacts related to the other unused impact elements of environmental stewardship, public infrastructure and healthy economy. However, because of the scope and intent of the assessment the most relevant and impactful indicators were used.*" – Intermediate

- Intermediate vs. Rapid – "the HIA in the Village of Weston to review local alcohol policies was performed as a "rapid" HIA over the course of *6 months.... The advisory committee for the Village of Weston HIA includes (stakeholders)* the Village Administrator; the Everest Metro Chief of Police, the Village Clerk and Marathon County Health Department Staff....The advisory committee has generated a list of *potential stakeholders (Village residents, Village Board members, license holders, youth, etc.) to contact to provide further information on existing conditions within the village and feedback* relating to the policy itself... Further research such as *focus groups, community surveys and stakeholder interviews* were considered in order to gather qualitative data from Village residents, business owners and leaders....." – Intermediate

Data Types – use the specified terminology, with the term "data" encompassing anything beyond models, literature, and websites (e.g., datasets, surveys, focus groups, consultations, interviews, etc.)

Impacts/Endpoints – use specified terminology for impacts assessed; any impacts that don't fit the specified terminology, enter "other"

Pathway of Impact – use specified terminology for the pathways or causes of the assessed impacts; details can be added in parentheses after the specified terminology [e.g., lifestyle (alcohol use, binge drinking, drunk driving, underage drinking)] or if any pathways don't fit the specified terminology, enter the name of the pathway

Quantification of Impact – use the specified terminology <u>and</u> identify the impacts; for likelihood, look for quantifiers such as likely, could, may, definitely when describing impacts; example:

Direction of impacts (positive impacts on health - increased opportunity for physical activity, improved safety, better access to health promoting goods and services, and enhanced social capital, as well as a slight reduction in car use and its negative health impacts; negative impacts on health - pedestrian and bicycle safety); likelihood of impacts (definite impacts - increased physical activity, enhanced social capital, better access; probable impacts - negative impacts related to pedestrian and bicycle safety; speculative impacts - reduction in car use and its negative health impacts)

Effectiveness of HIA – when describing the impact of the HIA on the policy, plan or project, use the terms direct, general, opportunistic, or none, along with a more detailed explanation; effectiveness can also include the accuracy of the HIA in predicting health impacts; *must be determined by an internet search*

3) Consistency in data entry (clarification of database fields)

Title – title of the HIA Report, exactly as it appears on the report cover (including any subtitles)

Decision-making Level – of policy, project, or plan being evaluated by the HIA (not of the HIA itself)

Organization(s) Involved – <u>lead</u> organization(s), conducting and publishing the HIA; not all organizations involved

Organization Type – type of organization <u>for each</u> of the organizations identified; *this will likely need to be identified via an internet search*

Contact – name and/or email (if available), otherwise enter "Undetermined"; no need to enter organization name, because entered in previous field

Organization/HIA Website – if there is not a website dedicated to or highlighting the HIA, enter "N/A"; no need to enter organization's general website if no info available on HIA; *this will likely need to be identified via an internet search*

Funding – only enter financial sponsors if specifically named in report (i.e., do not assume based on organizations involved); if unsure or not identified, enter "Undetermined"

Major Data Sources – major data sources are highlighted in the text and/or presented in tables, charts, or figures; name the specific organization/agency <u>and</u> data type, year, and geographic scale (if presented); examples: U.S.

Census Bureau 2009 (city- and county-level demographics and socio-economic status), New Mexico Department of Health (Hispanic rate of death and leading causes of death; city and county),

Local Data Available or Obtained? – name the local data obtained (e.g., health data; locations of vehicle accidents, vulnerable populations, etc.; zoning ordinances; demographics; income; life expectancy; transportation; population; crime; socioeconomic data; specific survey data; etc.)

Additional Data Needed (Self-Identified) – name the self-identified data needs

Stakeholder/Community Involvement? – identify the stakeholders involved AND avenue of involvement (interviews, focus groups, surveys, etc.)

Health Endpoints – name the general and/or specific health effects identified in the HIA (e.g., physical health, mental health, developmental health, asthma, cardiovascular disease, injury, fatality/mortality, chronic diseases, obesity/weight, infectious disease, headaches, malnutrition, etc.)

Impact Prioritization – describe how was it determined which pathways/impacts would be focused on (e.g., community input, equity/distribution of impacts, funding availability, synthesis of literature and data, magnitude of the impact, likelihood of the impact, permanence of the impact, etc.); if it is not clear from the report how this was done, enter "Undetermined" and if no prioritization took place (i.e., all impacts were assessed), enter "N/A; all impacts assessed")

Decision-making Outcome – the results reported in the HIA – conclusions, recommendations, mitigations, etc. - NOT the effect the HIA had on the plan, project, or decision; this should be entered in "Effectiveness of HIA" field

Defensibility/Process Evaluation – your evaluation of the quality of the process undertaken (evidence, methodology, assumptions, limitations, and barriers) and the documentation of that process, using the HIA practice standards (included in the Review Guidelines) as a benchmark; if authors documented a process evaluation, note this as well (e.g., identified successes, challenges, and lessons learned)

1. Quality of Evidence and Methodology
- Was the supporting information and methodology sound and clearly documented in the report (e.g., adequate literature, data etc. collected; sources of data acknowledged; clear description of data and methodology used; identification of participants and their roles, funding, etc.)?
- Was the scope of the HIA and process undertaken clearly documented?
- Was stakeholder input solicited and utilized?
- Were the recommendations based on transparent, context-specific synthesis of evidence (e.g., impacts/conclusions well supported by the data, literature, etc. presented in the report) or was it not clear how the authors reached the conclusions (e.g., evidence presented only spoke to general health impacts and not the specific impacts examined)?

2. Assumptions, Limitations, and Barriers
- Note any assumptions, limitations, and barriers identified in the report
- Identify assumptions, limitations, and barriers you saw in the HIA, that were not identified by the authors (e.g., data gaps, assumptions, etc.)

3. Documentation
- Is the documentation of the process, methods, findings, sponsors, participants, etc. transparent and publicly-accessible?

Follow-up Measures – monitoring, health impact management, or other follow-up measures called for in the HIA

Minimum Elements of HIA Met? If no, what's missing – If all aspects of the 5 elements (and sub-elements) are not met, the answer is no; enter "No;" and identify what elements or aspects of the elements are missing (e.g., No; Element 4 is missing and documentation of funding sources (Element 5) is not transparent)

GIS Used? – if yes, describe how GIS was used (e.g., illustrative maps of site locations, used in spatial analysis to evaluate proposed bike routes and existing traffic safety conditions)

4) Consistency in level of detail presented
   - Use pilot HIA reviews as a benchmark

   Scope/Summary
   Quantification of Impact
   Decision-making Outcome

5) Consistency in subjective evaluations
   - Review of an exemplary HIA (often one that involved Human Impact Partners, Robert E. Wood Foundation, or Health Impact Project) will provide a benchmark for subjective evaluations (Defensibility/Process Evaluation, Potential Improvements, Best Practices) and allows for consistency in these evaluations

II. Reviewer Feedback – Challenges, Questions, and Lessons Learned

- Primary factors in determining HIA rigor are the number of impacts assessed and level/depth of assessment – Duration of the HIA will vary depending on the number of people involved and the actual level of effort (PT or FT) and, therefore, is the least accurate criteria.

- Internet searches can be used to gather data for three fields only – Organization Type, Organization/HIA Website, and Effectiveness of HIA.

  Reason: One of the HIA standards is transparent documentation of the process, methods, findings, sponsors, funding, participants, etc. If this information is not clearly documented in the HIA Report, then we need this noted as such in order to accurately reflect the state of HIAs in these sectors and areas for improvement.

- Source of evidence, data types, and major data sources fields are used together to paint the data picture – Source of evidence is the method used to collect the data, data types is the general type of data used (models, literature, websites, and data), and major data sources defines the source and type of data [e.g., US Census Bureau 2009 (city- and county level demographics and socioeconomic status)].

- It is difficult to determine how to accurately categorize the HIAs (e.g., Decision-making Level, HIA Type, HIA Rigor, etc.) – In general, this will become easier as more HIAs are reviewed, but the HIAs are not always going to be black and white; in those cases, best judgement should be used in data entry and/or the specified terminology that best fits chosen.

- Typical health endpoints (morbidity/mortality) are not used and impacts are not quantified – The health endpoints in HIA are not typical toxicity study endpoints, although mortality (or fatality) may be an endpoint. Health endpoints are general and/or specific health effects, such as physical health, mental health, asthma, chronic disease, injury, obesity, malnutrition, etc.

  Impacts are not typically quantified as they are in scientific research, but by a more qualitative assessment. This can be due to a number of factors, including lack of available scientific research, unavailability of local data, time limitations, limited resources, etc.

- Determining the appropriate level of detail is difficult – Use the pilot HIA reviews as a benchmark, but in some cases, the depth of information provided in the HIA (for quantification of impacts or recommendations/mitigation, for example) makes it ineffective to enter information at that level of detail; in those cases, it is acceptable to summarize the information.

- Subjective evaluations of defensibility, potential improvements, and best practices are difficult – Review of an exemplary HIA (often one that involved Human Impact Partners, Robert E. Wood Foundation, or Health Impact Project) will provide a benchmark for subjective evaluations, but be sure to compare apples to apples – a rapid HIA will not have the same level of effort or detail that an intermediate or comprehensive HIA will have.

III. Review Pathforward

Next review deadline: July 13 - Half of the HIA reviews complete

- Given the discussion and the details provided in this document, it is acceptable (and likely warranted) to go back and revise data entry for the previous HIA reviews.

- For reference, each reviewer will be provided the primary and QA data entry for the HIAs they reviewed in the QA Review. Upon examination, if there are any questions, feel free to discuss with the other reviewer.

- One additional HIA will be selected in mid-July and reviewed by all five reviewers as part of a secondary QA Review.

- Toward the end of the HIA Review, reviewers will be asked to share their three most difficult HIA

Appendix D – Data Sources Used in Reviewed HIAs

Resource/Organization	Data Source	Description	Website
		Demographics and Background Info	
U.S. Census Bureau	American FactFinder	Provides access to data about the U.S., Puerto Rico, and the Island Areas from multiple U.S. Census Bureau censuses and surveys, including: Decennial Census, American Community Survey (ACS), American Housing Survey (AHS), Economic Census, Census of Governments, Population Estimates Program, and more.	http://factfinder2.census.gov/
	Decennial Census	Provides demographic, social, and economic data at state, county, city, zip code, census tract, block group, and block levels, every 10 years.	General information: http://www.census.gov/ Data: http://factfinder2.census.gov/
	American Community Survey (ACS)	An on-going survey that releases results each year. Instead of actual counts, it provides estimates based on a random sample of the population. It is used to collect data on demographic, social, and economic characteristics at state, county, and sometimes smaller levels (e.g., zip code tabulation area) depending on the year, for example: age, sex, race, family and relationships, income and benefits, health insurance, and education.	General information: http://www.census.gov/acs/www/ Data: http://factfinder2.census.gov/
	American Housing Survey (AHS)	A national housing sample survey that gathers information on the number and characteristics of U.S. housing units, as well as the households that occupy those units.	General information: http://www.census.gov/housing/ahs/ Data: http://factfinder2.census.gov/
	Economic Census	Provides a profile of national and local economies every five years.	General information: http://www.census.gov/econ/census/ Data: http://factfinder2.census.gov/

D-1

Resource/Organization	Data Source	Description	Website
Demographics and Background Info (Cont.)			
U.S. Department of Housing and Urban Development (HUD)	HUD USER	Provides access to Fair Market Rents data, Special Tabulations of Households, and many other original HUD datasets.	http://www.huduser.org/portal/pdrdatas_l anding.html
	Fair Market Rents	Gross rent estimates that include the shelter rent plus the cost of all tenant-paid utilities, except telephones, cable or satellite television service, and internet service. Used to determine how much rent should be covered through Section 8 for individuals with low income.	http://www.huduser.org/portal/datasets/f mr.html
	Special Tabulations of Households	Produces tabular statistical summaries of counts of households by income, tenure, age of householder, and housing conditions for select geographic areas in the U.S.	http://www.huduser.org/portal/datasets/sp ectabs.html
U.S. Department of Labor	Bureau of Labor Statistics (BLS)	Databases, tables, and calculators on essential economic information such as labor market activity (e.g., employment or unemployment), working conditions (e.g., pay and benefits), and price changes. Data are available at the state, county, and sometimes smaller geographic scales.	http://www.bls.gov/home.htm
Oregon Employment Department	Oregon Labor Market Information System	Provides statewide information on unemployment, employment by industry, wages, personal income and cost of living, consumer price index, and employer-provided benefits (e.g., health, retirement, leave, other), as well as regional economic and occupational profiles.* *similar labor statistics may be available for other states	http://www.qualityinfo.org/olmisi/Olmis Zine
Health Data			
U.S. Centers for Disease Control and Prevention (CDC)	Behavioral Risk Factor Surveillance System (BRFSS)	World's largest, on-going telephone health survey. This survey, which is run by CDC and conducted by individual state health departments, examines behavioral risk factors in the U.S.	http://www.cdc.gov/brfss/index.htm

Resource/Organization	Data Source	Description	Website
		Health Data (Cont.)	
U.S. Centers for Disease Control and Prevention (CDC)	National Center of Health Statistics (NCHS)	Provides access to data, documentation, and questionnaires for various national health surveys, such as the National Health Interview Survey (NHIS), National Health and Nutrition Examination Survey (NHANES), National Vital Statistics System (NVSS), and National Immunization Survey (NIS).	http://www.cdc.gov/nchs/
	National Health and Nutrition Examination Survey (NHANES)	A program of studies designed to assess the health and nutritional status of adults and children in the U.S. The survey is unique in that it combines interviews and physical examinations.	http://www.cdc.gov/nchs/nhanes.htm
	National Health Interview Survey (NHIS)	Data on a broad range of health topics are collected through personal household interviews. For over 50 years, the U.S. Census Bureau has been its data collection agent. Survey results have been instrumental in providing data to track health status, health care access, and progress toward achieving national health objectives.	http://www.cdc.gov/nchs/nhis.htm
	Youth Risk Behavior Surveillance System (YRBSS)	Monitors six types of health-risk behaviors that contribute to the leading causes of death and disability among youth and adults (i.e., behaviors that contribute to unintentional injuries and violence; sexual behaviors that contribute to unintended pregnancy and sexually transmitted diseases, including HIV infection; alcohol and other drug use; tobacco use; unhealthy dietary behaviors; and inadequate physical activity) and measures the prevalence of obesity and asthma among youth and young adults. Includes a national school-based survey conducted by CDC, and state, territorial, tribal, and local surveys conducted by state, territorial, and local education and health agencies and tribal governments.	http://www.cdc.gov/HealthyYouth/yrbs/index.htm

D-3

Resource/Organization	Data Source	Description	Website
		Health Data (Cont.)	
Georgia Department of Public Health	Online Analytical Statistical Information System (OASIS) / Health and Vital Statistics Data Repository	Provides access to the state's standardized health data repository*, which includes hospital discharge, emergency room visit, arboviral surveillance, YRBSS, BRFSS, sexually transmitted disease, motor vehicle crash, vital statistics (i.e., births, deaths, fetal deaths, induced terminations, pregnancies), and population data. Where possible, data are available by age group, race, ethnicity, sex, census tract, county commission district, county, health district, legislative district, region, or state. *similar health and vital statistics data may be available from other state, county, and local health departments*	http://oasis.state.ga.us/oasis/#
University of California at Los Angeles	California Health Interview Survey (CHIS) / State Health Survey	A state survey conducted every two years that provides key health statistics for adults, adolescents, and children. Data are available at the state, county, region, and service planning area levels in California.* *similar health survey data may be available for other states*	http://www.chis.ucla.edu/
Los Angeles County Department of Public Health	Los Angeles County Health Survey /County or City Health Survey	A periodic, population-based survey that provides information about the health of county residents on topics such as health outcomes, health behaviors, the built environment, and access to medical care. Data are available for Los Angeles County and its service planning areas and health districts.* *similar health survey data may be available for other counties and locales*	http://publichealth.lacounty.gov/ha/hasurveyintro.htm
		Other Supporting Data	
Denver Police Department	Crime Statistics	The Police Department provides crime statistics and maps at the city, neighborhood, police district, and city council district levels, and data on sex offenders, and gang activity.* *similar crime data may be available for other counties and locales*	http://www.denvergov.org/police/Police Department/tabid/440727/Default.aspx

D-4

Resource/Organization	Data Source	Description	Website
		Other Supporting Data (Cont.)	
Oregon Department of Environmental Quality (DEQ)	Environmental Databases and Mapping Applications	Provides access to environmental data, such as air quality, water quality, wastewater permits, enforcement actions, and cleanup sites, and provides GIS/mapping applications for capturing, managing, analyzing, and displaying the various geographically-referenced information.* *similar databases and applications may be available for other states and locales*	http://www.deq.state.or.us/news/databases.htm
U.S. Department of Agriculture (USDA)	Food and Nutrition Service	Provides access to various nutrition and hunger data, including data on food security, food assistance and nutrition programs, and Supplemental Nutrition Assistance Program (SNAP) and Summer Food Service Program (SFSP) participation rates and economic benefits.	http://www.fns.usda.gov/outreach/getinvolved/data.htm
Benton County Maps and GIS	Maps and GIS Repository	This GIS repository and mapping application* provides a user interface to view and query roads, parks, tax lot, zoning, survey documents, addressing, election maps, aerial photography, topography, and other digital map layers. *similar GIS data may be available for other states and locales*	http://www.co.benton.or.us/maps/bentonmaps.php
Oakland Parks and Recreation (OPR)	Parks, Recreation Facilities, and Programming	Provides locations of parks, recreation facilities, pools, etc. in the community, as well as information on programming.* *similar data on parks and recreation may be available for other states and locales*	http://www2.oaklandnet.com/Government/o/opr/index.htm
Clark County Assessor	Parcel Data and Property Records	Provides parcel data and maps, including data on roads and other right-of-way parcels, and a property search function, which allows users to access ownership and property value data.* *similar parcel data and property records may be available for other counties and locales*	http://www.clarkcountynv.gov/Depts/assessor/Pages/default.aspx

D-5

Resource/Organization	Data Source	Description	Website
		Other Supporting Data (Cont.)	
San Francisco Planning Department	Zoning, Permits, Planning Code	Provides the complete planning code, zoning and permit data, and survey maps.* *similar planning data may be available for other states and locales	http://www.sf-planning.org/
Los Angeles Unified School District	School Locator/School Profile & Performance	Provides a school locator, profiles of school demographics (e.g., enrollment, ethnicity, graduation rate, suspensions/expulsions, attendance rates), and school report cards (e.g., academic performance index, English and Math proficiency).* *similar databases and applications may be available for other states and locales	School Locator: http://notebook.lausd.net/schoolsearch/selector.jsp School Profile and Performance: http://data.lausd.net/why-does-data-matter-how-do-i-get-data
Federal Transit Administration (FTA)	National Transit Database	Provides monthly and annual financial and operating data on transit agencies throughout the U.S., including expenditures, revenue sources, service delivery, and trip length.	http://www.ntdprogram.gov/ntdprogram/
U.S. Department of Transportation, Federal Highway Administration	National Household Travel Survey (NHTS) [formerly Nationwide Personal Transportation Survey (NPTS)]	A national inventory of daily travel and includes information on the purpose and means of travel, travel time, day and time of travel, and traveler demographics.	http://nhts.ornl.gov/introduction.shtml
California Highway Patrol	Statewide Integrated Traffic Records System (SWITRS)	A database of California collision data, including bicyclist and pedestrian collisions.* Custom reports are available by criteria, such as jurisdiction, location, or date. *similar databases may be available for other states and locales	http://www.chp.ca.gov/switrs/index_menu.html

Resource/Organization	Data Source	Description	Website
		Other Supporting Data (Cont.)	
Oregon Department of Transportation (DOT)	Transportation Data Section	Provides statewide transportation data, such as pavement condition, transportation infrastructure, traffic counts and flow, posted speed limits, traffic congestion, transit routes, vehicle miles travelled, and crash data, as well as mapping applications for capturing, managing, analyzing, and displaying the various geographically-referenced information.* *similar databases and applications may be available for other states and locales*	http://www.oregon.gov/ODOT/TD/TDA TA/Pages/index.aspx ; https://gis.odot.state.or.us/transgis/
Metro Transit	Transit Services	provides maps of transit routes, as well as service frequency, average ridership, and fare data.* *similar transit data may be available for other regions and locales*	http://www.metrotransit.org/
		Benchmarks	
U.S. Centers for Disease Control and Prevention (CDC)	Physical Activity Guidelines for Americans	Identify how much physical activity children, adults, older adults, and healthy pregnant/ postpartum women should be getting and provide examples of different types of activities to meet those goals.	http://www.cdc.gov/physicalactivity/ever yone/guidelines/index.html
U.S. Environmental Protection Agency (EPA)	Air Quality/Pollution Standards	Under the Clean Air Act, EPA is responsible for setting standards, also known as national ambient air quality standards (NAAQS), for pollutants which are considered harmful to people and the environment and ensuring that these air quality standards are met, or attained through national standards and strategies to control pollutant emissions from automobiles, factories, and other sources. Under the Clean Air Act, federal noise regulations have also been set, which cover standards for transportation equipment, motor carriers, low-noise-emission products, and construction equipment. Although EPA is not responsible for regulating indoor air quality, as it does outdoor air quality, the Agency does conduct indoor air quality research to examine the health risks of radon, mold, and other indoor air pollutants and offer means by which to reduce human exposures.	National Ambient Air Quality Standards: http://www.epa.gov/air/criteria.html Air Pollutants /Air Quality: http://www.epa.gov/air/airpollutants.html Transportation-related Air Quality Standards: http://www.epa.gov/otaq/ Noise Pollution: http://www.epa.gov/air/noise.html Indoor Air Quality: http://www.epa.gov/iaq/index.html

Resource/Organization	Data Source	Description	Website
Benchmarks (Cont.)			
Robert Wood Johnson Foundation and the University of Wisconsin Population Health Institute	County Health Rankings & Roadmaps	Using data available for each county in all 50 states, measure the overall health of each county based on factors that influence health, such as health behaviors, clinical care, social and economic factors, and the physical environment. Users can see specific county-level data and state benchmarks for various measures used in the rankings.	http://www.countyhealthrankings.org/
Transportation Research Board / National Cooperative Highway Research Program	Crash Reduction Factors	This State-of-Knowledge Report summarizes the current status of crash reduction factors for a variety of treatments and provides a summary of the best available crash reduction factors.	http://onlinepubs.trb.org/onlinepubs/nchrp/nchrp_rrd_299.pdf
National Crime Prevention Council	Crime Prevention Through Environmental Design (CPTED)	A multi-disciplinary approach to deterring criminal behavior through environmental design implementations in the built environment.	http://www.ncpc.gov.sg/pdf/CPTED%20Guidebook.pdf
World Health Organization (WHO)	Health Initiatives, Strategies, and Guidelines	Initiatives and evidenced-based strategies and guidelines for various health and development topics, including community noise.	http://www.who.int/topics/en/ ; http://www.who.int/publications/guidelines/en/index.html
U.S. Department of Health and Human Services	Healthy People	A set of science-based, ten-year national health objectives.	Healthy People 2010: http://www.healthypeople.gov/2010/ Healthy People 2020: http://www.healthypeople.gov/2020/default.aspx
Indicators			
San Francisco Department of Public Health	Sustainable Communities Index (SCI); formerly Healthy Development Measurement Tool (HDMT)	A set of measurement methods for indicators of livable, equitable, and prosperous cities. Includes over 100 measures that can be used to track diverse sustainability objectives for the environment, transportation systems, community cohesion and civic engagement, public facilities, education, housing, and economic strength, and health systems. Where possible, methods try to represent indicators at the neighborhood scale.	http://www.sustainablesf.org/

D-8

Appendix E – Tools and Models Used in Reviewed HIAs

Tool/Model	Description	Source
20-Minute Neighborhood Analysis	A GIS analysis of walkability and local access to services that takes into account both the presence of local destinations, as well as factors that impact the ability to access these destinations (e.g., street connectivity, sidewalks, transit service, and topography).	Portland Bureau of Planning and Sustainability; http://efiles.portlandoregon.gov/webdrawer.dll/webdrawer/rec/4376218/view/PP%2020-min%20neighborhood%20analysis.PDF
AERMOD	A dispersion model used to estimate criteria pollutants.	American Meteorological Survey/Environmental Protection Agency; http://www.epa.gov/ttn/scram/dispersion_prefrec.htm#aermod
Air Quality Index (AQI) Scores/AirData	The AirData website allows users to display and download monitored hourly, daily, and annual concentration data, AQI data, and particle pollution data collected at outdoor monitors across the U.S. Puerto Rico, and the U.S. Virgin Islands. The data come primarily from the Air Quality System (AQS) database.	Environmental Protection Agency; http://www.epa.gov/airdata/
Bicycle Environmental Quality Index (BEQI)	A quantitative observational survey used to assess the bicycle environment on roadways and evaluate what streetscape improvements could be made to promote bicycling.	San Francisco Department of Public Health; http://www.sfphes.org/component/jdownloads/viewcategory/19-beqi?Itemid=62
CALINE3/CAL3QHC/CAL3QHCR	CALINE3/CAL3QHC are pollutant dispersion models used for predicting carbon monoxide (CO) dispersion from traffic. CAL3QHCR is a more refined version that requires local meteorological data. Inputs for the model include: roadway geometry; receptor locations, meteorological conditions, and vehicular emission rates.	Environmental Protection Agency; http://www.epa.gov/ttn/scram/dispersion_prefrec.htm#caline3; http://www.epa.gov/ttn/scram/dispersion_prefrec.htm#cal3qhc

Tool/Model	Description	Source
CALINE4 (CAlifornia LINE Source Dispersion Model)	Based on the same diffusion equation used in EPA's CALINE3 model, CALINE4 is the standard modeling program used by the California Department of Transportation to predict air concentrations of CO near roadways; the model can also handle dispersion modeling of particulate matter (PM) and nitrogen dioxide (NO2). Inputs for the model include traffic characteristics (volumes, speeds, etc.), roadway geometry, meteorological data, and vehicle emission factors.	California Department of Transportation (Caltrans); http://www.dot.ca.gov/hq/env/air/software/caline4/calinesw.htm
CALRoads View	An air dispersion modeling package that combines the CALINE4, CAL3QHC, and CAL3QHCR air dispersion models into one seamless integrated graphical interface. This package is used for predicting air pollution concentrations of CO, NO2, and PM from traffic.	Lakes Environmental Software; http://www.weblakes.com/products/calroads/2.AspxAutoDetectCookieSupport=1
Carbon Sequestration Estimates	A methodology for determining possible sequestration ability of local forest cover.	U.S. Department of Agriculture; http://www.fs.fed.us/ecosystemservices/pdf/estimates-forest-types.pdf
Childcare Supply and Demand	A methodology for determining childcare supply and demand in a community, and can also be used to determine childcare demand created by new development.	Enterprise Community Partners; http://www.practitionerresources.org/showdoc.html?id=19705&topic=Resident%20Services&doctype=Manual
Claritas BusinessPoint (now Nielsen BusinessPoint)	An on-line data source that matches and appends large business databases in real time to provide detailed statistics including sales, number of employees, primary contacts, existing markets, and market potential.	http://www.claritas.com/MyBestMarkets2/Default.jsp?ID=0&SubID=&pageName=Home
Comprehensive Plan Review Checklist	Designed for use in reviewing comprehensive land use plans, transportation plans, and neighborhood plans, and is appropriate for different locations. Elements examined include land use, transportation, water resources, parks and open space, and urbanization/redevelopment/ economic development.	Design for Health; http://designforhealth.net/resources/legacy/checklists/
Consumer Price Index	A tool that provides monthly data on changes in the prices paid by urban consumers for a representative basket of goods and services.	U.S. Bureau of Labor Statistics; http://www.bls.gov/cpi/

Tool/Model	Description	Source
Diversity Index	A proprietary diversity index that measures diversity on a scale from 0 (no diversity) to 100 (complete diversity). Defined as the likelihood that two persons, selected at random from the same area, would belong to a different race or ethnic group.	ESRI; http://www.esri.com/data/esri_data/demographic-overview/demographic (Note: The Diversity Index is part of ESRI's 2012/2017 Updated Demographics database.)
Emissions & Generation Resource Integrated Database (eGRIDweb)	A web-based tool that displays eGRID data in a user friendly way and allows users to export data they select; a comprehensive source of data on the environmental characteristics of almost all electric power generated in the U.S. and links air emissions data with electric generation data for U.S. power plants.	Environmental Protection Agency; http://cfpub.epa.gov/egridweb/
EMission FACtor Model (EMFAC)	An emission inventory model used to calculate emission rates from motor vehicles, such as passenger cars to heavy-duty trucks, operating on highways, freeways, and local roads in California; the emission rates are multiplied with vehicle activity data provided by regional transportation agencies to calculate the statewide or regional emission inventories. Inputs required for generating an emissions inventory are geographic area, calendar year, month or season selection, title, model years included in the calculation, inspection and maintenance (I/M) programs, emission mode, and output options.	California Air Resources Board; http://www.arb.ca.gov/msei/onroad/latest_version.htm
Food Access Research Atlas (formerly Food Desert Locator)	An internet-based mapping tool that provides a spatial overview of food access indicators for low-income and other census tracts using different measures of supermarket accessibility and pinpoints the location of food deserts (low-income communities that lack ready access to healthy food).	U.S. Department of Agriculture; http://www.ers.usda.gov/data-products/food-access-research-atlas.aspx
Frank and Sallis GIS-Based Walkability Index	A combined measure of net residential density, road network connectivity, retail floor area ratio, and land use mix, and can be calculated using archival GIS data rather than by means of intensive primary data collection efforts.	Frank L, Sallis J, Saelens B, Leary L, Cain K, Conway T, et al. 2010. The development of a walkability index: application to the neighborhood quality of life study. *British Journal of Sports Medicine* 43:924-933.

Tool/Model	Description	Source
Geographic Information Systems (GIS) / Mapping Applications	GIS is a tool that allows users to visualize, analyze, interpret, and understand geographically-referenced data to reveal relationships, patterns, and trends. Examples include ArcGIS (a GIS for working with maps and geographic information), SimplyMap (a web-based mapping application with a user-friendly interface), and MapInfo (a desktop-based GIS used for mapping and location analysis).	ESRI; http://www.esri.com/software/arcgis Geographic Research, Inc.; http://geographicresearch.com/simplymap/ Pitney Bowes; http://www.mapinfo.com/
Google Map/ Google Earth	Google Maps is a web mapping application that offers street maps, a route planner for traveling by foot, car, bike, or with public transportation, and a locator for urban businesses in numerous countries around the world. Google Earth combines maps and geographic information with satellite and aerial photography, allowing the user to view 3D imagery, terrain, and buildings. Note that Earth View is a feature in Google Maps that allows users to see the same high-resolution imagery, terrain, and 3D buildings that are available in the desktop version of Google Earth.	Google; http://maps.google.com/; http://www.google.com/earth/index.html
Greenhouse Gas (GHG) Emissions Modeling	The Puget Sound GHG Emissions Model evaluates various GHG models, identifies the most appropriate model for the project under consideration, and identifies appropriate values for key factors and components of the scenarios being analyzed.	Puget Sound Clear Air Agency; SR520 HIA, Appendix K - http://www.wsdot.wa.gov/NR/rdonlyres/EFDE4CC6-406F-48E4-BEFD-EF50B2842625/0/SR520HealthImpactAssessment.pdf
Health Economic Assessment Tool (HEAT)	A tool designed to conduct an economic assessment of the health benefits of walking or cycling by estimating the value of reduced mortality that results from regular walking or cycling.	World Health Organization; http://www.heatwalkingcycling.org/

Tool/Model	Description	Source
Healthy Development Management Tool (HDMT); now Sustainable Communities Index (SCI)	A comprehensive set of measurement methods for indicators of livable, equitable, and prosperous cities. Includes over 100 measures that can be used to track sustainability objectives for the environment, transportation systems, community cohesion and civic engagement, public facilities, education, housing, and economic strength, and health systems and apply these metrics to planning, policy making, and civic engagement.	San Francisco Department of Public Health; http://www.sustainablesf.org/
Health Impact Predictive Function Equations	Used to predict excess traffic-attributable PM2.5 mortality, changes in noise-related annoyance, and noise-attributable cases of myocardial infarction.	San Francisco Department of Public Health; Road Pricing HIA, Appendices B & C - http://www.sfphes.org/component/jdownloads/finish/37-congestion-pricing/111-health-effects-of-road-pricing-in-san-francisco-california/0?Itemid=0
HIA Threshold Analysis	A detailed spreadsheet-based assessment that uses a point-based scoring system to assess achievement across a wide variety of planning-related topics, including accessibility, air quality, environment and housing quality, food, mental health, physical activity, safety, social capital, and water.	Design for Health; http://designforhealth.net/wp-content/uploads/2012/12/BCBS_HIAThreshold4.0_063008.pdf
Home Energy Insecurity Scale	A tool used to quantitatively measure the extent to which an energy assistance program improves the energy self-sufficiency of a low-income household.	U.S. Department of Health and Human Services; http://www.acf.hhs.gov/sites/default/files/ocs/measuring_outcome_0.pdf
Huff Gravity Model (integrated into GIS)	A probabilistic retail gravity model (created by D. Huff 1963) used to predict consumer behavior among competing retail shopping locations. From these probabilities, sales potential can be calculated based on disposable income, population, or other variables.	ESRI; ArcGIS 9.3: http://arcscripts.esri.com/details.asp?dbid=15999; ArcGIS 10.0 or later: http://www.arcgis.com/home/item.html?id=f4769668fc3f486a992955ce55caca18
Input-Output Model	Shows the value of goods and services flowing among the various economic sectors. This model provides a detailed and complete picture of a state or region's economic structure, including inter-industry linkages, and the economy's dependence on different markets.	State of Hawaii*; http://files.hawaii.gov/dbedt/economic/data_reports/2005_state_io/2005-input-output-study.pdf *similar models may be available for other states and areas

Tool/Model	Description	Source
Land Use Regression Model	Utilizes monitored levels of the pollutant of interest as the dependent variable and variables such as traffic, topography, and other geographic variables as the independent variables in a multivariate regression model to characterize air pollution exposure and health effects that vary spatially.	General information on land use regression models: http://www.integrated-assessment.eu/guidebook/land_use_regression ; Review of land use regression models: http://www.ncbi.nlm.nih.gov/pmc/articles/PMC2233947/pdf/nihms3668 3.pdf
Living Wage Calculator	Estimates the cost of living in a community or region. Lists typical expenses, the living wage, and typical wages for the selected location.	Poverty in America Project (Penn State); http://livingwage.mit.edu/
Location Allocation Model (MINDISTANCE)	A site analysis technique in GIS using location allocation to determine potential locations for retail that optimize proximity. Specifically computes the location and allocation to minimize the total weighted distance traveled from all demand points to their nearest center.	ESRI; ArcGIS Network Analyst extension ArcGIS 9.3: http://webhelp.esri.com/arcgisdesktop/9.3/index.cfm?TopicName=Creating an OD cost matrix ArcGIS 10.0 or later: http://help.arcgis.com/en/arcgisdesktop/10.0/help/index.html#//0047000000500000000
Log-linear Risk Model of Population Exposure - Particulate Matter	Forecasts the mortality effects of exposure to particulate matter for a population.	World Health Organization; http://www.who.int/publications/cra/chapters/volume2/1353-1434.pdf
Mapping Susceptibility to Gentrification: Early Warning Toolkit	Developed to help communities in California identify whether their neighborhood is susceptible to gentrification. Provides indicators to identify neighborhoods at risk of gentrification.	U.C. Berkley Center of Community Innovations; http://communityinnovation.berkeley.edu/reports/Gentrification-Report.pdf
Metropolitan Sprawl Index	Measures and evaluates metropolitan sprawl based on four factors: residential density; neighborhood mix of homes, jobs, and services; strength of activity centers and downtowns; and accessibility of the street network.	Smart Growth America; http://www.smartgrowthamerica.org/research/measuring-sprawl-and-its-impact/
Neighborhood Environment Walkability Scale (NEWS)	A survey that assesses residents' perception of neighborhood design features related to physical activity, including residential density, land use mix (including both indices of proximity and accessibility), street connectivity, infrastructure for walking/cycling, neighborhood aesthetics, traffic and crime safety, and neighborhood satisfaction.	http://sallis.ucsd.edu/measure_news.html

Tool/Model	Description	Source
Noise Annoyance Relationship	Formulas used to define populations a little annoyed, annoyed, or highly annoyed by aircraft, road traffic, and railway noise.	http://www.ncbi.nlm.nih.gov/pmc/articles/PMC1240282/pdf/ehp0109-000409.pdf Miedema HME and Oudshoorn CGM. 2001. Annoyance from transportation noise: relationships with exposure metrics DNL and DENL and their confidence intervals. *Environmental Health Perspectives* 109(4):409-416.
Pedestrian Environment Data Scan (PEDS)	An audit instrument that measures environmental features related to walking in varied environments in the U.S.	http://planningandactivity.unc.edu/RP1.htm
Pedestrian Environmental Quality Index (PEQI)	A quantitative observational tool used to assess the pedestrian environment. The tool is organized into five categories: intersection safety, traffic, street design, land use, and perceived safety. Indicators within these categories are aggregated to create a weighted summary index, which can be reported as an overall index score.	San Francisco Department of Public Health; http://peqiwalksafe.com/
Primer on Gentrification and Policy Choices	A primer on how to view the complex issue of gentrification, including nationally-recognized indicators to measure whether gentrification is occurring.	Brookings Institution Center on Urban and Metropolitan Policy/ PolicyLink; http://www.policylink.org/atf/cf/%7B97C6D565-BB43-406D-A6D5-CA3BBF35AF0%7D/DealingWithGentrification_final.pdf
Retail Food Environment Index (RFEI)	A tool that gives a snapshot of unhealthy versus healthy food retail options for an area. RFEI is a ratio of fast food restaurants & convenience stores, divided by the number of groceries, farmer's markets, and produce stands; a higher RFEI suggests greater concentration of unhealthy food.	California Center for Public Health Advocacy; http://www.publichealthadvocacy.org/RFEI/presskit_RFEI.pdf
Retail Gap Analysis	A technique for identifying the strengths and weaknesses of an economy's retail sector. The technique quantifies retail surplus and leakage (i.e., the extent to which the corridor is capturing the spending potential of households residing in the area).	http://pods.dasnr.okstate.edu/docushare/dsweb/Get/Document-1631/F-917web.pdf
SF-CHAMP Travel Forecasting Model	A regional travel demand model that is used assess estimated changes in travel patterns in the San Francisco Bay area under different land use, population, and transportation system conditions.	San Francisco County Transportation Authority; http://www.sfcta.org/modeling-and-travel-forecasting

Tool/Model	Description	Source
Sleep Disturbance Formula	Calculates the percent of the exposed population expected to be awakened by single event noise exposure.	U.S. Federal Interagency Committee on Noise; http://www.fican.org/pdf/nai-8-92.pdf
Smart Growth Parking Demand Model	Estimates how many parking spaces are required in a study area based on its demographics, current land uses, and projected developments. Also helps tailor parking requirements by land use for each study area, incorporating a series of case study-specific factors such as alternative transportation mode prevalence, the ability to share parking, and the time period under study.	Metropolitan Transportation Commission (San Francisco Bay); http://www.mtc.ca.gov/planning/smart_growth/parking/parking_seminar.htm
SoundPLAN Noise Model	Noise modeling software that can predict, analyze, and graphically display traffic noise, occupational noise indoors and outdoors, general industrial noise, and aircraft noise. SoundPLAN is a standards-based program that can provide road noise calculations in accordance with various international standards, including the Federal Highway Administration's STAMINA/Traffic Noise Model.	SoundPLAN International LLC and Braunstein + Berndt GmbH; http://www.soundplan.eu/english
Store Inventory Tool	Used to inventory the availability and pricing of fresh fruit and vegetables, snack foods, and beverages, and assess store infrastructure.	D.C. Healthy Corner Store Program; http://www.dchunger.org/pdf/cornerstores08_phaseone_report.pdf; Appendix A
Student Generation Rates for New Residential Development	Can be used to project public school students, by level (elementary, middle, high), from proposed residential development.	Oakfield Unified School District*; http://www.ousd.k12.ca.us/cms/lib07/CA01001176/Centricity/Domain/95/Oakland%20USD%20-Developer%20Fees%20Study.pdf *similar information may be available for other states and school districts
Traffic Congestion Burden Index	Quantifies the combined effect of congestion and the degree to which people are exposed to it. The index considers measures of rush-hour traffic and travel rates with figures for the portion of commuters who are subject to that congestion because they drive to work.	Surface Transportation Policy Project; http://www.transact.org/pdfs/etb_report.pdf

E-8

Tool/Model	Description	Source
Traffic Noise Model (TNM; formerly STAMINA)	A Windows-based model for predicting and analyzing highway traffic noise. Inputs for the model include data on the roadway, receivers, barriers, building rows, terrain lines, ground zones, tree zones, contour zones, etc. This model replaces the previous STAMINA 2.0/OPTIMA Model.	Federal Highway Administration; http://www.fhwa.dot.gov/environment/noise/traffic_noise_model/
Urban Emissions Model (URBEMIS)	Estimates air pollution emissions in pounds per day or tons per year for various land uses, area sources, construction projects, and project operations. Note that URBEMIS 2007 uses California motor vehicle emission rates, which tend to be lower than those in other states due to California's stricter emission controls; therefore, out-of-state users should adjust emission outputs to reflect their vehicle fleets.)	California Air Resources Board; http://www.urbemis.com/
Vehicle-Cyclist Injury Collision Predictive Model Equation	Used to estimate future vehicle-cyclist injury collisions and % change in vehicle-cyclist collisions. Model inputs are the number of motor vehicles and the number of cyclists in the different study scenarios.	Elvik R. 2009. The non-linearity of risk and the promotion of environmentally sustainable transport. *Accident Analysis and Prevention* 41(4):849-855
Vehicle-Pedestrian Injury Collision Model	A census-tract level model of pedestrian injury collision frequency as a function of aggregate traffic volume (log-transformed), street, land use and population characteristics. The model can be used to estimate census-tract level changes in vehicle-pedestrian injury collisions.	San Francisco Department of Public Health; http://www.sfphes.org/elements/24-elements/tools/108-pedestrian-injury-model Wier M, Weintraub J, Humphreys EH, Seto E, Bhatia R. 2009. An area-level model of vehicle-pedestrian injury collisions with implications for land use and transportation planning. *Accident Analysis and Prevention* 41(1):137-145
Walk Score	A public access walkability index that assigns a numerical walkability score to any address in the U.S., Canada, or Australia. The Walk Score is a number between 0 (car dependent) and 100 (walker's paradise).	Walk Score; http://www.walkscore.com/

Appendix F – Identified Data Gaps and Additional Data Needs

Note: The HIAs included in this Appendix were found to have self-identified additional data needs.

ID	HIA Title	Location	Additional Data Needs (Self-Identified)
1	Pathways to a Healthy Decatur: A Rapid Health Assessment of the City of Decatur Community Transportation Plan	Decatur, Georgia	Yes; city-scale health data (county data used instead)
2	Affordable Housing and Child Health: A Child Health Impact Assessment of the Massachusetts Rental Voucher Program	Massachusetts	Yes; Department of Community and Housing Development data on program participants and program utilization
3	Health Impact Assessment - Derby Redevelopment, Historic Commerce City, Colorado	Historic Commerce City, Colorado	Yes; sufficient research to identify the relative importance of the community design features that promote physical activity
4	Health Impact Assessment of NRMT's Request for a Special Use Permit	Bernalillo County, New Mexico	Yes; more complete/consistent applicant information (data on types, numbers, and age of fleet vehicles; information on type of waste transport and waste transport routes; waste volume, waste origin, and waste characterization; consistent traffic projections)
5	The Red Line Transit Project Health Impact Assessment	Baltimore, Maryland	Yes; locally-placed air monitors to assess air pollution (greater air pollution predicted in communities in the Red Line corridor as of result of increased traffic with the No Build Option)
6	Health Impact Assessment Report: Alcohol Environment - Village of Weston, WI	Village of Weston, Wisconsin	Yes; consistently-reported health behavior data for Wisconsin youth; local data on youth alcohol use and effects of underage drinking
7	Eastern Neighborhoods Community Health Impact Assessment Final Report	San Francisco, California	Yes; locate more qualitative info to supplement and inform the quantitative data collected; data gaps in healthy economy profile - psycho-social attributes of jobs (physical work conditions, job security, access to health insurance through employment, lack of control over work, lack of participation in decision making, time spent at work, supportive work environment, work-life balance); data on the quality of public services and infrastructure
8	Health Impact Assessment: South Lincoln Homes, Denver CO	Denver, Colorado	Yes; a study on children and adolescent health; vehicle-to-vehicle and vehicle-to-pedestrian accident data
9	Spokane University District Pedestrian/Bicycle Bridge Health Impact Assessment	Spokane, Washington	Yes; physical activity data for the quarter-mile radius around the bridge
12	Community Health Assessment: Bernal Heights Preschool - An Application of the Healthy Development Measurement Tool (HDMT)	San Francisco, California	Yes; parental, social, and environmental factors affecting selection of childcare and school locations (e.g., commute times, etc.); adequate information available to apply the HDMT development target checklist
13	St. Louis Park Comprehensive Plan - Health Impact Assessment	St. Louis Park, Minnesota	Yes; identification of pollutant sources in the neighborhood

ID	HIA Title	Location	Additional Data Needs (Self-Identified)
14	Comprehensive Health Impact Assessment: Clark County Bicycle and Pedestrian Master Plan	Clark County, Washington	Yes; qualitative data on existing bicycle and pedestrian infrastructure; comprehensive inventory of pedestrian facilities; local health data (morbidity/mortality) linked to built environment data; data on most types of morbidity by neighborhood; data on physical activity by neighborhood; data on racial/ethnic disparities (due to small numbers); poverty data at a small geographic scale; overweight/obesity data by zip code
16	Yellowstone County/City of Billings Growth Policy Health Impact Assessment	Yellowstone County, Montana	Yes; supplementary research to show causality between elements of the built environment and chronic disease
17	Health Impact Assessment, June 20, 2011: Duluth, Minnesota's Complete Streets Resolution, Mobility in the Hillside Neighborhoods and the Schematic Redesign of Sixth Avenue East	Duluth, Minnesota	Yes; noise assessment
19	Alaska Outer Continental Shelf - Beaufort Sea and Chukchi Sea Planning Areas, Oil and Gas Lease Sales 209, 212, 217, and 221 Draft Environmental Impact Statement; Appendix J - Public Health	Alaska	Yes; baseline health and environmental data
21	Fort McPherson Rapid Health Impact Assessment: Zoning for Health Benefit to Surrounding Communities During Interim Use	Atlanta, Georgia	Yes; data on current health status of communities surrounding Fort McPherson
22	Re: November 10th Merced County General Plan Update (MCGPU) Preferred Growth Alternative Decision	Merced County, California	Yes; additional data and analysis would be required in order to conduct a full HIA; a mechanism for predicting potential health impacts of proposed land use and policy decisions
23	Health Impact Assessment of Modifications to the Trenton Farmer's Market (Trenton, New Jersey)	Trenton, New Jersey	Yes; more data on the proximate impacts of markets and more locale-specific data on the prevalence of pertinent risk factors is needed for more quantitative analysis of farmers market health impacts
26	Health Impact Assessment: Hawai'i County Agriculture Development Plan	Hawai'i County, Hawaii	Yes; availability of detailed, accurate information on the expenditures of the School Food Services Branch to make sound recommendations for increasing procurement of local food
27	Mass Transit Health Impact Assessment: Potential Health Impacts of the Governor's Proposed Redirection of California State Transportation Spillover Funds	California	Yes; actual impacts of funding cutbacks on transit services

ID	HIA Title	Location	Additional Data Needs (Self-Identified)
31	Health Impact Assessment for Proposed Coal Mine at Wishbone Hill, Matanuska-Susitna Borough Alaska	Matanuska-Susitna Borough, Alaska	Yes; Social Determinants of Health - current household level survey data for the potentially affected communities and data for years 2009-2011; Accidents and Injuries - data for years 2009-2011; Exposure to Potentially Hazardous Materials - offsite residential monitoring well data, updated fish/aquatics data set, site specific PM2.5 data, air permit application, analysis of potential dust/diesel emissions, visual effects analysis, and off-site terrain noise modeling; Food, Nutrition and Subsistence - Traditional and Local Knowledge survey and updated subsistence data/analysis
33	Neenah-Menasha Sewerage Commission Biosolids Storage Facility, Greenville, WI	Greenville, Wisconsin	Yes; risks of groundwater contamination; community (citizen) surveys (as study relied on unstructured public comment and literature about community opinions and perceptions)
35	A Health Impact Assessment of Accessory Dwelling Unit Policies in Rural Benton County, Oregon	Benton County, Oregon	Yes; more current data than the 2000 and 2006 census data used; data on unincorporated areas; ADU literature for rural areas; quantitative data on number of permits to be requested; quantitative data to replace the qualitative data collected at community meeting and advisory panel meetings
36	The Health Impact Assessment (HIA) of the Commonwealth Edison (ComEd) Advanced Metering Infrastructure (AMI) Deployment	Northern Illinois (Commonwealth Edison Utility Territory)	Yes; data to estimate impact to customer bills
37	Atlanta Beltline Health Impact Assessment	Atlanta, Georgia	Yes; information about park and trail design, entry points, and changes to the surrounding environment to allow a more accurate assessment of access
38	Zoning for a Healthy Baltimore: A Health Impact Assessment of the Transform Baltimore Zoning Code Rewrite	Baltimore, Maryland	Yes; literature that takes socioeconomic status into account when looking at the health impacts of the built environment; longitudinal studies or randomized controlled trials that further delineate the relationship between the built environment and health
39	Hood River County Health Department Health Impact Assessment for the Barrett Property	Hood River County, Oregon	Yes; soil testing results indicating whether pesticide chemical residues exist in the soil
40	Unhealthy Consequences: Energy Costs and Child Health - A Child Health Impact Assessment of Energy Costs and the Low Income Home Energy Assistance Program	Boston, Massachusetts	Yes; utility company data on arrearages and shut-offs in Massachusetts
46	The Sellwood Bridge Project: A Health Impact Assessment	Multnomah County, Oregon	Yes; bike and pedestrian safety data; details on construction equipment to be used to allow for air pollution and noise impacts during construction to be assessed
48	A Rapid Health Impact Assessment of the City of Los Angeles' Proposed University of Southern California Specific Plan	Los Angeles, California	Yes; precise data on displacement; qualitative data of neighborhood changes to identify communities receiving displaced persons

ID	HIA Title	Location	Additional Data Needs (Self-Identified)
49	Taylor Energy Center Health Impact Assessment	Taylor County, Florida	Yes; epidemiological data on the impact of climate change; health impact data for ozone exposure
54	A Health Impact Assessment on Policies Reducing Vehicle Miles Traveled in Oregon Metropolitan Areas	Oregon [Portland, Eugene-Springfield, Rogue Valley (Medford-Ashland area), Corvallis, Bend, and Salem-Keizer]	Yes; baseline data for vulnerable populations
55	Health Effects of Road Pricing In San Francisco, California: Findings from a Health Impact Assessment	San Francisco, California	Yes; air quality measures other than PM2.5
56	Santa Monica Airport Health Impact Assessment	Santa Monica, California	Yes; demographics
59	Douglas County Comprehensive Plan Update Health Impact Assessment	Douglas County, Minnesota	Yes; water quality data; additional mapping to fully understand the County's bike network and recreational amenities and their connections to residential areas and other services; 2010 census data for updated population map
60	The Executive Park Subarea Plan Health Impact Assessment: An Application of the Healthy Development Measurement Tool (HDMT)	San Francisco, California	Yes; more detailed baseline health data (baseline conditions used mostly demographic data, less health data); more specific implementation strategies; more details to support Sustainable and Safe Transportation and Public Infrastructure and Healthy Economy elements
67	Healthy Tumalo Community Plan: A Health Impact Assessment on the Tumalo Community Plan; A Chapter Of The 20-Year Deschutes County Comprehensive Plan Update	Tumalo, Oregon	Yes; more research in the areas of nearby recreation, trail system development, and health outcomes in both urban and rural settings; further evaluation of impacts of short-term ODOT solution to safe highway accessibility
68	Strategic Health Impact Assessment on Wind Energy Development in Oregon (Public Review Draft)	Oregon	Yes; epidemiological studies on sound, shadow flicker, amplitude modulation, and indoor low frequency sound impacts; accident rates due to driver distraction; and wind energy impacts on jobs, income, and other economic indicators
69	Impacts on Community Health of Area Plans for the Mission, East SoMa, and Potrero Hill/Showplace Square: An Application of the Healthy Development Measurement Tool	San Francisco, California	Yes; neighborhood-level data on gross per capita water usage, annual per capita waste disposal, tree canopy, proportion jobs paying self-sufficiency wage and filled by residents, proportion households living on income below self-sufficiency standard, occupational injury, jobs providing health insurance, proportion locally-owned businesses, planned parking pricing strategies, planned traffic calming interventions, public transit access to public school, proportion children attending neighborhood schools, access to produce stores and food markets, neighborhood completeness indicator for key public and retail services, volunteerism, sidewalks with adequate lighting
70	Pathways to Community Health: Evaluating the Healthfulness of Affordable Housing Opportunity Sites Along the San Pablo Avenue Corridor Using Health Impact Assessment	Oakland, California	Yes; mortality and morbidity data for El Cerrito

ID	HIA Title	Location	Additional Data Needs (Self-Identified)
71	MacArthur BART Transit Village Health Impact Assessment	Oakland, California	Yes; information to assess compliance with existing housing law; maintenance and repair of MBTV; cost of the project housing; retail effects; effectiveness of particular interventions for reducing pedestrian injuries
72	Healthy Corridor for All: A Community Health Impact Assessment of Transit-oriented Development Policy in St. Paul Minnesota	St. Paul, Minnesota	Yes; agriculture, forestry, fishing, hunting, mining, quarrying, and oil and gas extraction industry average annual wages; income at the block level
73	Health Impact Assessment Point Thomson Project	Alaska	Yes; data related to human consumption of subsistence resources to accurately assess the affects of nutrition changes
77	Humboldt County General Plan Update Health Impact Assessment	Humboldt County, California	Yes; demographic and resources data below the zip code level
78	Rapid Health Impact Assessment: Vancouver Comprehensive Growth Management Plan 2011	Vancouver, Washington	Yes; qualitative data on existing bicycle and pedestrian infrastructure; comprehensive inventory of pedestrian facilities; record-level local health data (morbidity/mortality) linked to built environment data; data on most types of morbidity by neighborhood; data on physical activity by neighborhood; data on some racial/ethnic disparities (due to small numbers)
81	Health Impact Assessment of the Port of Oakland	Oakland, California	Yes; a more comprehensive study of truck counts and activity; quality of pedestrian environment (PEQI); quality of parks and open space; estimation of social cohesion and level of physical activity; eligibility requirements for port employment and degree to which those positions are fulfilled by West Oakland residents

Table G-1. Summary Table of Select Data from HIAs in Each Sector

ID	HIA; Year; Location	Scope/Summary	Source of Evidence	Impacts/ Endpoints	Pathways of Impact	Characterization of Impact	Decision-making Outcome	Effectiveness of HIA	Minimum Elements of HIA Met? If no, what's missing
1	Pathways to a Healthy Decatur: A Rapid Health Assessment of the City of Decatur Community Transportation Plan; 2007; Decatur, Georgia	Examine the health impacts of the City of Decatur Community Transportation Plan that aims to make Decatur a healthy place to live and work, maintain a high quality of life, and increase opportunities for alternative modes of transportation	Community consultation, literature review, special collection (expert consultation, demographics analysis, GIS)	Health, behavioral, infrastructure	Land use, lifestyle, mobility/access to services, physical activity, safety (traffic) and security, social capital	Direction of impacts (positive health impacts, negative health impacts); Likelihood of impacts (definite impacts, probable impacts, speculative impacts)	Concluded that elements of the Community Transportation Plan, such as intersection and corridor improvements, bike and pedestrian facilities, and transportation and land use connections, will increase opportunities for physical activity, improve safety, and provide better access to health promoting goods and services. Potential negative health impacts exist related to pedestrian and bicycle safety, but can be eliminated or mitigated by incorporating findings of the HIA in the design phase of the corridor and intersection improvements. Recommendations included: making traffic safety a priority, prioritizing connectivity, making intersections ADA-compliant, emphasizing the mobility of the most vulnerable populations, supporting the bicycle community, partnering with schools to promote childhood physical activity, accommodating commuter and recreation users in planning alternate transportation modes, a community-wide campaign to promote physical activity, and making the Plan one component of a greater health promoting strategy for the city	General effectiveness assumed at a minimum – HIA was included as an Appendix to the Community Transportation Plan and portions incorporated into the body of the Plan; infra-structure improvements and an Active Living Division created, but it is unclear if this is a direct result of health recommendations in the HIA being implemented	No; Element 4 (monitoring plan) missing; permanence of potential impacts on human health or health determinants (Element 2.4) not judged; and documentation of funding (Element 5) not transparent
5	The Red Line Transit Project Health Impact Assessment; 2008; Baltimore, Maryland	Examine current health conditions for the population living in the Red Line corridor, illustrate links between transportation and health in Baltimore, and recommend specific design features and mitigation strategies to maximize the Baltimore Red Line Project's capacity to achieve better health	Literature review, community consultation, policy review, special collection (interviews, expert consultation, modeling health links, demographics analysis)	Health, behavioral, environmental, infrastructure, services, demographic	Air quality, exposure to hazards, land use, mental health, mobility/access to services, noise pollution, nutrition, physical activity, safety and security, social capital	Direction of impacts (positive health impacts, negative health impacts); Equity of impacts (demographics, populations sensitive to vehicle emissions and noise); Magnitude of impacts (high for some risks); Likelihood of impacts (definite positive impact, speculative positive impacts)	Potential for some negative health impacts from construction, but the majority of health impacts are positive; the No-Build option eliminates the potential for the benefits the Red Line offers. Cross-cutting Recommendations: Build the Red Line using light rail, appoint a public health expert to serve on decision-making teams, and increase green space. Recommendations were also provided for improving access and safe outdoor activity and construction issues	General effectiveness at a minimum, although direct effectiveness possible – HIA was submitted to the Maryland Transit Authority as comment to the draft EIS; in 2009, Maryland Governor announced locally-preferred alternative for the Red Line would be implemented – a light rail system	No; Element 4 (monitoring plan) missing; documentation of funding not transparent (Element 5)

Table G-1. Continued

ID	HIA, Year; Location	Scope/Summary	Source of Evidence	Impacts/ Endpoints	Pathway of Impact	Characterization of Impact	Decision-making Outcome	Effectiveness of HIA	Minimum Elements of HIA Met? If no, what's missing
9	*Spokane University District Pedestrian/ Bicycle Bridge Health Impact Assessment*; 2011; Spokane, Washington	Inform decision makers about potential health impacts that development of a pedestrian bridge in the University District will have on the current and projected population who will live, work, and recreate within a quarter-mile radius of the bridge	Literature review, policy review, special collection (surveys, demographics analysis; GIS)	Health, environmental	Safety and security, social capital, air quality, physical activity, community economics, housing, land use	Correlative analyses of eyes on the streets and windows on the block to perceived safety and perceived safety or eyes on the street with actual crime numbers; Direction of impacts (positive health impacts, negative health impacts); Magnitude of impacts (hypothesize a positive impact on physical activity similar in magnitude to London study); Likelihood of impacts (unlikely impacts, likely impacts); Cost of impacts (pedestrian and bicycle collisions)	The bridge will contribute positively to the health of the study area and the HIA recommend that the bridge be constructed A number of recommendations were produced and prioritized for implementation, including reduced on- and off-street parking, incentives for alternative transportation and mixed-use development, bike lanes on and to/from the bridge, regular bus service, proper repair and maintenance of sidewalks, maps and signage for bicycle and pedestrian routes, traffic calming measures, and continued branding of the University District	Undetermined	No; Element 4 (monitoring plan) missing; and distribution and permanence of potential impacts not assessed (Element 2 4)
11	*The Impact of U.S. Highway 550 Design on Health and Safety in Cuba, New Mexico: A Health Impact Assessment*; 2010; Cuba, New Mexico	Provide information on how the design of US Highway 550 could impact the health and safety of Cuba area residents and visitors	Literature review, community consultation, special collection (survey, workshop)	Health, behavioral, infrastructure, economic	Land use, safety and security, physical activity, social capital, community economics, mobility/access to services	Draws on general research about the effects of transportation planning on pedestrian safety, physical activity, social connections, and community economics, but doesn't touch on specific health consequences or quantify impacts for Cuba residents	Recommendations were provided that could increase pedestrian safety and encourage safe walking along US 550 as part of an active daily lifestyle for Cuba area residents, including improving the pedestrian environment (adding or upgrading sidewalks; providing lighting, shade, and benches; and creating safe pedestrian crossings); providing a buffer between vehicular traffic and foot traffic; signage that would promote a safe pedestrian environment; and traffic calming measures	General effectiveness at a minimum, although direct effectiveness possible – per the Winter 2011 UNMPRC Newsletter, the HIA was included in the NMDOT's Environmental Assessment of the Cuba sidewalk project	No; Elements 2 2 (stakeholder input), 2 4 (judgment of impacts), and 4 (monitoring plan) missing; evidence of the HIA being initiated to inform a decision-making process is not clear (Element 1); documentation of the process and methods not very transparent (Element 5)

Table G-1. Continued

ID	HIA; Year; Location	Scope/Summary	Source of Evidence	Impacts/ Endpoints	Pathway of Impact	Characterization of Impact	Decision-making Outcome	Effectiveness of HIA	Minimum Elements of HIA Met? If no, what's missing
14	*Comprehensive Health Impact Assessment: Clark County Bicycle and Pedestrian Master Plan*; 2010; Clark County, Washington	Examine the likely health impacts of the Clark County Bicycle and Pedestrian Master Plan, whether to adopt the Master Plan or not, and how elements of the Plan could be prioritized to maximize health impacts	Literature review, special collection (expert consultation, demographics analysis, GIS)	Health, infrastructure, economic	Physical activity primarily; to a limited extent: mobility/access to services, parks and recreation, nutrition, community/household economics, safety	Projects - Direction of impacts (positive health impacts); Strength of evidence (limited, some, moderate, or strong); Magnitude of impacts (populations served based on GIS analysis); Distribution/ equity of impacts (one-way ANOVA to determine disparate impacts); Permanence of impacts (medium – medical treatment for obesity); Quantification of impacts (medical costs of obesity, correlations between fast food density and income, bicycle network density); Policies/Programs – Direction of impact; Strength of evidence; Distribution/equity of impacts (if applicable)	Concluded that the projects, policies, and programs in the Bicycle and Pedestrian Master Plan would positively impact health by increasing opportunities for physical activity The HIA identified important data inputs omitted during the planning process, which limited the ability of the plan to maximize health benefits Overarching recommendations included: updating the plan in five years; using data to prioritize projects and track progress; and planning and providing for the needs of a continuum of users and trip types Project, policy, and program recommendations were also provided	Direct effectiveness -- the board of County Commissioners adopted the plan, which will be incorporated into the 2014 County Comprehensive Growth Management Plan Update	Yes, although permanence of impacts (Element 2 4) is not very apparent
17	*Health Impact Assessment, June 20, 2011: Duluth, Minnesota's Complete Streets Resolution, Mobility in the Hillside Neighborhoods and the Schematic Redesign of Sixth Avenue East*; 2011; Duluth, Minnesota	The purpose of the HIA was to determine the potential health impacts of the Sixth Avenue East Schematic Redesign Study, if the redesign was embracing Duluth's Complete Streets Resolution, and how the redesign study could be improved to provide additional health benefits to users of the corridor	Literature review, policy review, community consultation, special collection (expert consultation, walkability audit, GIS)	Health, infrastructure, behavioral	Mobility/access to services, safety and security, physical activity, livability, community economics, housing, social capital, land use, parks and recreation	Distribution/equity of impacts (transit mobility for vulnerable populations); otherwise, impacts not really documented	Concluded that the roadway redesign is a feasible project, that the Comprehensive Plan has additional language to support the redesign, and with existing city policies/plans and the recommendations made in the HIA that the renewal of the corridor would positively impact health and better serve the users of and residents of Hillside Selected recommendations addressed accessibility and safety, physical activity, and livability Suggested that recommendations become an addendum to the Redesign Study and be paired with an upcoming traffic study	Undetermined	No; Elements 2 3 (baseline health conditions) and 2 4 (judgment of impacts) missing

G-3

Table G-1. Continued

ID	HIA; Year; Location	Scope/Summary	Source of Evidence	Impacts/ Endpoints	Pathway of Impact	Characterization of Impact	Decision-making Outcome	Effectiveness of HIA	Minimum Elements of HIA Met? If no, what's missing
27	*Mass Transit Health Impact Assessment: Potential Health Impacts of the Governor's Proposed Redirection of California State Transportation Spillover Funds;* 2008; California	Synthesize and communicate research evidence on how proposed cuts in state funding of mass transit may impact the public's health and inform pending transportation funding decisions in California and illustrate how public policies outside the public health and health care sectors can affect public health	Literature review, policy review, special collection (consultation with transit advocacy group, expert consultation, modeling)	Health, economic, environmental, other (social)	Air quality, water quality, noise pollution, community/ house-hold economics, land use, physical activity, lifestyle, social capital, mobility/access to services	Due to uncertainties, could not predict health impacts of state transit funding cutbacks; provided pathways through which transportation and transit funding generally affect public health	Concluded that getting people out of their cars and into mass transit has the potential to benefit health in a number of ways Uncertainty about how the state transit funding cutbacks would affect transit systems throughout the state and manifest at the local level made it difficult to draw firm conclusions about the health impacts However, the HIA predicted cuts were most likely to impact smaller agencies that lack other resources to make up the funds and for transit-dependent populations, such as the children, seniors, low-income and disabled persons	General effectiveness – the budget for fiscal year 2007, which included the Governor's proposed re-direction of $1.3 billion in transportation "spillover" funds to the State's General Funds, was approved prior to the HIA Report being issued, but allocation of public funds for transit at the state and local levels continues to be a high priority issue	No; Elements 2 4 (judgement of impacts), 3 (recommendations and mitigations), and 4 (monitoring plan) missing; stakeholder input limited (Element 2 2); although initiated in advance of the decision, was not completed in advance of the decision (Element 1)
32	*HIA of the Still/Lyell Freeway Channel in the Excelsior District;* 2008; San Francisco, California	Examine the health impacts associated with past construction of the I-280 Freeway and high-traffic surface streets in the Excelsior District of San Francisco after concerns surfaced that residents of that community were being disproportionately exposed to traffic-related exposures, including air pollution and suffering the health consequences	Literature review, community consultation, special collection (survey, interviews, community photography, traffic counts, modeling, walkability audit, demographics analysis, GIS)	Health, environmental	Air quality, noise pollution, safety and security	Quantification of impacts (air and noise pollution impacts via modeling, assessment of the Pedestrian Environment using the Pedestrian Environmental Quality Index); Direction of impacts (negative health impacts of traffic)	Air quality and noise modeling and monitoring provided evidence that traffic contributed significantly to environmental hazards in the Excelsior neighborhood, which is largely composed of families with children, immigrants, and people of color Also found that leading causes of death in the project zip code were illnesses associated with increased exposure to traffic and traffic-related air pollutants and noise, including heart disease, lung cancer, and traffic collisions The HIA identified solutions to the risks identified, such as using more non-polluting (hybrid) buses, reducing truck traffic, building a sound wall next to the freeway, establishing a program for acoustic upgrades to building facades and windows, ensuring safe routes to school, and improving health care access	Direct effectiveness – lobbied San Francisco Board of Supervisors to draft and adopt a resolution to reduce the adverse health impacts of local truck traffic on southeast communities, and on November 25, 2008, the Board unanimously passed Resolution 081397, which requires a mitigation plan to address the impacts of local truck traffic on residential communities of southeast San Francisco	No; Elements 1 (conducted in advance of decision) and 4 (monitoring plan) are missing; documentation of funding sources and HIA point-of-contact not transparent and HIA documentation is scattered and not easily accessible (Element 5); magnitude, likelihood, distribution, and permanence of impacts not addressed (Element 2 4)

Table G-1. Continued

ID	HIA; Year; Location	Scope/Summary	Source of Evidence	Impacts/ Endpoints	Pathway of Impact	Characterization of Impact	Decision-making Outcome	Effectiveness of HIA	Minimum Elements of HIA Met? If no, what's missing
42	Columbia River Crossing Health Impact Assessment; 2008; Multnomah County, Portland, Oregon	Examine the Columbia River Crossing (CRC) Draft Environmental Impact Statement (EIS) through a public health lens to understand the scope and magnitude of the potential health effects of the four bridge alternatives being considered	Literature review, policy review	Health, environmental/ ecosystem, behavioral, economic, infrastructure, services	Air quality, exposure to hazards, mental health, mobility/ access to services, noise pollution, physical activity, safety and security	Direction of impacts (positive health impacts, negative health impacts); Distribution/equity of impacts (demographics, populations sensitive to noise, transit mobility for vulnerable populations)	Provided six recommendations and suggested additional analyses in several areas, including transportation safety, air quality, noise pollution, and environmental justice The HIA and health-based recommendations were submitted as a detailed comment letter during the public comment period for the draft EIS Recommendations included use of light rail, transit alignments that serve low income and minority populations, roadway interchange improvements, safe and accessible bike and pedestrian facilities, tolling to discourage single occupancy motor vehicle use, and alternatives that do not increas single occupancy motor vehicle use Suggestions for additional analyses included travel forecasting and predicted collision rates, analysis of cumulative air toxics exposure, analysis of noise impacts using a lower threshold, etc	Undetermined – bridge project is still under consideration	No; Elements 2 2 (stakeholder input), 2 3 (baseline health conditions), and 4 (monitoring plan) missing; magnitude, likelihood, and permanence of impacts not assessed (Element 2 4); assumptions and limitations not acknowledged (Element 2 5); documentation of funding sources and HIA point-of-contact not transparent (Element 5)
46	The Sellwood Bridge Project: A Health Impact Assessment; 2011; Multnomah County, Oregon	Assess how the proposed Sellwood Bridge redesign may affect human health during both the construction and operational phases of the project	Literature review, special collection (GIS)	Health, behavioral, infrastructure	Air quality, exposure to hazards, mental health, noise pollution, physical activity, safety and security	Direction of impacts (positive health impacts, negative health impacts); Magnitude of impacts (population estimates for air and noise pollution impacts); Likelihood of impacts (likely impacts, definite impacts); Distribution/equity of impacts (increased risk for air and noise pollution impacts based on proximity to the bridge; populations sensitive to noise and air quality impacts)	Concluded that the replacement for the current Sellwood Bridge is expected to be beneficial to county residents' health by increasing safety and opportunities for bicyclists and pedestrians, while addressing general transportation concerns The HIA recommended additional measures to maximize safety along the corridor and maintain safe air quality and noise levels during construction In addition, the HIA identifies potential partners for implementation of those measures	Undetermined	No; Elements 2 2 (stakeholder input) and 4 (monitoring plan) missing; permanence of impacts not assessed (Element 2 4); scoping phase is not clear (Element 2 1); and documentation of funding sources not transparent (Element 5)

Table G-1. Continued

ID	HIA; Year; Location	Scope/Summary	Source of Evidence	Impacts/ Endpoints	Pathway of Impact	Characterization of Impact	Decision-making Outcome	Effectiveness of HIA	Minimum Elements of HIA Met? If no, what's missing
51	*Rapid Health Impact Assessment, Crook County/City of Prineville, Bicycle and Pedestrian Safety Plan*; 2011; Crook County, Oregon	Evaluate the current pedestrian and bicycle situation in Prineville, Oregon and provide recommendations to be incorporated into the updated Bicycle and Pedestrian Safety Plan	Literature review, community consultation, policy review, special collection (demographics analysis, windshield survey, GIS)	Health, behavioral, infrastructure, services	Exposure to hazards, lifestyle, mental health, mobility/access to services, noise pollution, nutrition, physical activity, safety and security	Direction of impacts from current conditions (negative health impacts, positive health impacts); Direction of recommendation impacts (positive health impacts); Likelihood of recommendation impacts (speculative impacts); Magnitude of recommendation impacts (affected populations identified); Distribution/equity of current conditions and recommendation impacts (populations vulnerable to transit mobility)	Resulted in the identification of some potential negative health impacts that could be eliminated or mitigated by incorporating the findings and results of the HIA into future planning. Recommendations were provided for increasing opportunities for physical activity, improving safety, and providing better access to health promoting goods and services. These included: increasing connectivity of existing sidewalks, increasing overall existence of sidewalks, maintaining and upgrading existing bike lanes, increasing overall amount of bike lanes, strategically reviewing speed limit zones, and creating safe pedestrian crossing in key traffic areas	Undetermined	Yes
53	*SR 520 Health Impact Assessment: A Bridge to a Healthier Community*; 2008; King County, Washington	Ensure health consequences were considered in the decision-making process for the SR 520 Bridge Replacement and HOV Project and help decision makers evaluate the alternatives based upon their potential health effects	Literature review, policy review, special collection (Mediation Group, modeling, GIS)	Health, environmental/ ecosystem, behavioral, economic, infrastructure, services	Air quality, community/house-hold economics, education, exposure to hazards, healthcare access, lifestyle, mental health, mobility/ access to services, noise pollution, physical activity, safety and security, social capital, water quality, land use	Direction of impacts (negative health impacts, positive health effects); Distribution/equity of impacts (higher concentrations of certain air pollutants for individuals in proximity to SR 520); Quantification of impacts (greenhouse gas analysis conducted)	Indicated that choosing the right set of features for the SR 520 Project – regardless of which of the three plans under consideration is adopted – could contribute significantly to improving the health of people in communities adjacent to the corridor and the livability of their neighborhoods. Recommendations (that would be useful in any alternative) were organized into the following critical health elements: construction period (reduced construction pollution, increased traffic management, noise control); transit, bicycling and walking (improved transit service, connected walking and bicycling facilities, way-finding system); landscaped lids and green spaces (landscaped freeway lids, improved and preserved green space, preserved access to the waterfront); and design features (noise reduction, additions to the visual character with art and design, stormwater management practices)	Opportunistic effectiveness – SR 520 Bridge Replacement and HOV Project would be implemented; this HIA ensured public health was adequately addressed in the decision-making process	No; Element 4 (monitoring plan) missing; permanence, magnitude, and likelihood of impacts not assessed (Element 2.4); and documentation of funding sources not transparent (Element 5)

Table G-1. Continued

ID	HIA; Year; Location	Scope/Summary	Source of Evidence	Impacts/ Endpoints	Pathway of Impact	Characterization of Impact	Decision-making Outcome	Effectiveness of HIA	Minimum Elements of HIA Met? If no, what's missing
54	A *Health Impact Assessment on Policies Reducing Vehicle Miles Traveled in Oregon Metropolitan Areas*; 2009; Oregon [Portland, Eugene-Springfield, Rogue Valley (Medford-Ashland area), Corvallis, Bend, and Salem-Keizer]	Assess how vehicle miles travelled (VMT) reduction strategies being considered by Oregon's six metropolitan regions would bring about changes in air quality, physical activity, and car accident rates—and what impact that would have on the public's health	Literature review, policy review, special collection (advisory committee)	Health, behavioral, environmental/ ecosystem, infrastructure, services	Air quality; exposure to hazards, lifestyle; mobility/access to services; physical activity, safety and security, community/ household economics	Direction of impacts (negative health impacts, positive health impacts); Magnitude of impacts (high, moderate, low, negligible, or insufficient); Equity of impacts (effects on vulnerable populations); Quality of evidence	Demonstrated that reducing VMT would have significant health benefits overall. Examined 11 different policies that could reduce VMT and recommended the 5 that would be the most beneficial in terms of the public's well-being. The HIA recommended a combination of improvements to the built environment (e.g., mixed-use and highly-dense with good connectivity, pedestrian and bicycle infrastructure improvements, traffic calming measures, air infiltration systems in buildings), increased costs (e.g., fees for employee parking at businesses in metropolitan areas, a VMT tax, income tax bracket-based fees/taxes), and strengthening of public transit (e.g., increased transit coverage, public transit with lower levels of area-specific pollution such as light rail). While all three policies work best together to reduce adverse health effects, built environment and strengthening public transit were found to be more positive policies for vulnerable populations	Undetermined – HIA was presented to seven government decision-making bodies; while the impact of the HIA is unclear, the final Jobs and Transportation Bill of 2009 and a subsequent bill included VMT targets for the six metropolitan areas in Oregon	No; Element 4 (monitoring plan) missing (suggests 'Future research,' but not specific monitoring measures); and permanence and likelihood of impacts not assessed (Element 2.4)

Table G-1. Continued

ID	HIA, Year, Location	Scope/Summary	Source of Evidence	Impacts/ Endpoints	Pathway of Impact	Characterization of Impact	Decision-making Outcome	Effectiveness of HIA	Minimum Elements of HIA Met? If no, what's missing
55	*Health Effects of Road Pricing In San Francisco, California: Findings from a Health Impact Assessment*, 2011; San Francisco, California	Examine a future road pricing scenario being studied by the San Francisco County Transportation Authority (SFCTA) that would charge $3 during AM/PM rush hours to travel into or out of the northeast quadrant of San Francisco (which includes a concentration of San Francisco's currently congested downtown streets)	Literature review, community consultation, policy review, special collection (demographics analysis, walkability audit, modeling, GIS)	Health, environmental/ ecosystem, behavioral, economic, infrastructure, services	Air quality, community/house-hold economics, exposure to hazards, lifestyle, mental health, mobility/access to services, noise pollution, physical activity, safety and security	Direction of impacts (negative health impacts, positive health impacts); Permanence (severity) of impacts (high, low); Magnitude of impacts (low, low to medium, substantial); Likelihood of impacts (certain/probable impacts, probable/speculative impacts); Distribution/equity of impacts (vulnerable populations identified; equity of transportation health effects spatially; potential for policy to reduce some inequitable adverse health effects); Quantification of impacts (estimates of PM2.5 traffic-related deaths, noise-related annoyances, and myocardial infarction; Health Economic Assessment Tool; area-level vehicle- pedestrian injury forecasting; street/ intersection conditions using Pedestrian Environment Quality Index; quantified changes in vehicle-cyclist injury collisions; economic value of traffic-attributable beneficial and adverse health effects); Overall confidence in impact assessment (uncertainty factors regarding magnitude)	Concluded that transportation system operation in San Francisco has highly significant health burdens and benefits today and that health burdens are expected to increase due to increased motor vehicle traffic and population densities; Road pricing could moderate, but not entirely eliminate, the expected health burdens associated with "business as usual," including increased populations and traffic and no new policies or funding to manage the transportation system; Recommendations were developed to enhance the potential health benefits of road pricing, support reductions in transportation- associated health costs, and increase active transportation and included: increasing congestion pricing fees in circumstances likely to result in reduced health risks (e.g., on "spare the air" days or applying specifically to more polluting vehicles); investing in walking and biking safety improvements in locations where injuries are greatest (e.g., with traffic calming along arterials in and near the road-pricing zone); using quieter, low-emission hybrid buses in areas where noise and air pollution are worse; investing in walking and biking infrastructure to encourage trips by foot and by bike into and out of the road pricing zone; monitoring road-pricing implementation to address any unanticipated traffic increases and health impacts; encouraging active transportation and discouraging driving through more policies such as demand-based parking fees, "unbundling" parking in new development, and transportation demand management programs	Undetermined	Yes

Table G-1. Continued

ID	HIA; Year; Location	Scope/Summary	Source of Evidence	Impacts/ Endpoints	Pathway of Impact	Characterization of Impact	Decision-making Outcome	Effectiveness of HIA	Minimum Elements of HIA Met? If no, what's missing
56	*Santa Monica Airport Health Impact Assessment*; 2010; Santa Monica, California	Organize, analyze, and evaluate existing information and evidence regarding Santa Monica Airport's (SMO's) impact on three issue areas: lack of an airport buffer zone, noise, and air quality	Literature review, policy review, special collection (interviews, expert consultation)	Health, environmental/ ecosystem	Air quality, exposure to hazards, land use, noise pollution	Direction of impacts (negative health impacts)	Offered feasible recommendations that could be taken into consideration to mitigate the adverse health impacts the airport's operations have on the surrounding communities, including eliminating or significantly decreasing the number of jet takeoffs, installing HEPA filters in surrounding schools and residential homes, implementing additional noise abatement strategies, notifying residents and affected community members of noise and air pollution risks, and maintaining a runway buffer zone. Closure of SMO would eliminate all health risks associated with airport and noise pollution	General effectiveness – HIA raised awareness	No: Elements 2.2 (stakeholder involvement), 2.3 (baseline health conditions) and 4 (monitoring plan) missing; scoping phase did not include impacts other than health (Element 2.1); magnitude, likelihood, distribution, and permanence of impacts not assessed (Element 2.4); and documentation of HIA point-of-contact not transparent (Element 5)
62	*Health Impact Assessment on Transportation Policies in the Eugene Climate and Energy Action Plan*; 2010; Eugene, Oregon	Examine the positive and negative impacts of transportation policies within the Eugene Climate and Energy Action Plan (CEAP). The HIA examined seven transportation objectives/recommend-ations and summarized the scientific evidence that links these policies to health issues in Eugene	Literature review, community consultation, policy review, special collection (demographics analysis, modeling)	Health, environmental/ ecosystem, behavioral, infrastructure	Air quality, safety and security, physical activity	Direction of impacts (positive health impacts, negative health impacts); Likelihood of impacts (speculative impacts)	Concluded that the Transportation and Land Use objectives of the CEAP have broad benefits and should be approved. Also provided recommendations to maximize positive impacts and mitigate negative impacts associated with the Plan, such as strategies to reduce greenhouse gas emissions and mitigate negative health impacts from increased urban density; investments in complete streets, safety improvements, and public transit; and systems to track injuries and fatalities by transportation mode, to evaluate plan implementation, and systematically improve bicycle and pedestrian outcomes. Also recommended incorporating HIA practices into transportation and land use planning at the state and local level	Direct effectiveness – on September 15th, 2010, Eugene's City Council unanimously endorsed Eugene's first Community Climate and Energy Action Plan	No: judgment of magnitude, distribution, and permanence of impacts not assessed (Element 2.4)
65	*Health Impact Assessment (HIA) of Proposed "Road Diet" and Restriping Project on Daniel Morgan Avenue in Spartanburg, South Carolina*; 2012; Spartanburg, South Carolina	Assess what the expected effect of the proposed Daniel Morgan Avenue (DMA) Road Diet and Restriping Project would be on the safety of motorists, bicyclists, and pedestrians; opportunities for physical activity; opportunities for improved access to goods and services; and air quality	Literature review, policy review, special collection (demographics analysis, GIS)	Health, environmental, infrastructure, services	Safety and security, physical activity, mobility/access to goods and services, air quality	Direction of impacts (positive health impacts); Magnitude of impacts (high, medium, low); Significance of impacts (high, medium, low); Likelihood of impacts (uncertain impacts, unlikely impacts, possible impacts, likely impacts, very likely impacts); Distribution/ equity of impacts (whole community, disproportionate effects)	Findings suggested that the proposed road diet and re-striping could not only improve the health of many people, but also prevent death, injury, and/or serious illnesses. Recommended that the City of Spartanburg implement the proposed road diet and re-striping and gave particulars (turning 4-lane road into 3-lane road, shared-use paths, wider sidewalks, physically-seperated bike lanes, etc.); also recommended expansion and marketing of existing bicycle lending program, ample road safety signs, implementing an educational program on rules of the road, and offering a community cycling safety class	Undetermined	Yes

Table G-1. Continued

ID	HIA; Year; Location	Scope/Summary	Source of Evidence	Impacts/ Endpoints	Pathway of Impact	Characterization of Impact	Decision-making Outcome	Effectiveness of HIA	Minimum Elements of HIA Met? If no, what's missing
66	*Treasure Island Community Transportation Plan*; 2009; San Francisco, California	Evaluated whether the Treasure Island Transport-ation Plan met the health needs of its neighborhood residents, using the HDMT assessment tool and focused on ways the transportation system could be designed and implemented to maximize opportunities for active modes of transport-ation - such as walking and cycling - and minimize the risk of injuries	Special collection (modeling, HDMT)	Health, economic	Physical activity, safety and security, mobility/access to services, community economics	Quantification of impacts (Healthy Development Measurement Tool Sustainable and Safe Transportation element development objectives, street and intersection conduction using Pedestrian Environmental Quality Index/ Bicycle Environmental Quality Index); Direction of impacts (positive health impact, negative health impacts)	Analysis indicated that the Transportation Plan met 13 of the 20 development targets in the Sustainable and Safe Transportation element of the HDMT Recommendations from the HDMT analysis included the possible construction of a pedestrian and bicycle connection, a policy regarding pedestrian improvements at locations with potential high frequencies of pedestrian collisions, a policy to address economic barriers to public transit utilization, elimination of residential parking requirements, a reduction in number of residential parking spaces per unit, targeted areas for traffic calming, and more detail regarding potential parking pricing strategies in final version of the plan	Direct effectiveness – some HIA recommendations included in the Treasure Island Transportation Implementation Plan issued in 2011	No; Elements 2 1 (scoping), 2 2 (stakeholder input), 2 3 (baseline health conditions), and 5 (transparent documentation) missing from the portion deemed the HIA; and likelihood, permanence, magnitude, and distribution of impacts not assessed (Element 2 4); some of these elements were addressed in the overall Transportation Plan, however
75	*Interstate 75 Focus Area Study Health Impact Assessment*; 2010; Cincinnati, Ohio	Review the final recommendations of the Revive Cincinnati: Neighborhoods of the Mill Creek Valley Comprehensive Plan and assess the health impacts of proposed Interstate-75 infrastructure improvements to select neighborhoods adjacent to I-75 Due to time constraints, this HIA only examined health impacts to two of the four focus areas in the study	Literature review, special collection (demographics analysis, air quality study, GIS)	Health, environmental, other	Air quality, mental health, safety and security, housing, mobility/access to services	Direction of impacts (negative health impacts, positive health impacts); Likelihood of impacts (probable impacts, definite impacts); Distribution/equity of impact (empowerment zone neighborhoods identified for assessment)	Concluded that several recommendations from the Comprehensive Plan would have positive impacts for air quality and public health, but concluded that the I-75 infrastructure improvements would most likely have a negative impact on air quality, specifically particulate matter 2 5 (PM2 5) and volatile organic compounds (VOCs) One overarching recommendation was to conduct an air quality study (focused on PM2 5 and VOCs) that included establishment of baseline air quality levels prior to construction, and air quality monitoring during and after construction Other recommendations touched on air quality, traffic/crashes, displacement, pre-construction, construction, and post-construction	Undetermined, but general effectiveness assumed – agreement for air quality monitoring established with the University of Cincinnati and baseline monitoring initiated; work on the HIA raised awareness among the Traffic and Engineering Department to solicit comments on several other transportation projects Lessons learned and experience from this HIA gave the Health Department the confidence to undertake 3 additional HIAs in Cincinnati	No; Elements 2 2 (stakeholder involvement) and 2 3 (baseline health conditions) missing; magnitude and permanence of impacts not assessed (Element 2 4); portions of Element 2 5 (assumptions, strengths and limitations) weak or missing, and documentation of funding sources not transparent (Element 5)
79	*Lake Oswego to Portland Transit Project: Health Impact Assessment*; 2010; Portland, Oregon	Complement the Draft Environmental Impact Statement (DEIS) and more fully assess the health impacts of the three transit alternatives of the Lake Oswego to Portland Transit Project- no-build, enhanced bus service, and streetcar	Literature review, special collection (modeling/ forecasting GIS)	Health, environmental, services	Air quality, physical activity, mobility/access to services, parks and recreation safety and security	Based on selection of build scenario versus no-build scenario; Direction of impacts (positive health impacts, negative health impacts, no change in health); Likelihood of impacts (definite impacts, probable impacts); Distribution/ equity of impacts (no disproportionate impacts)	Provided recommendations for mitigating the adverse health impacts of either build scenario - enhanced bus service or streetcar Recommendations included: developing more stringent emissions-based fleet requirements or incentives; improving construction equipment emissions; outreach to residents regarding construction and potential health effects; education on how to avoid exposure to air toxics generated during construction; and monitoring programs to assess construction site concentrations of air toxics	Undetermined – the Lake Oswego to Portland Transit Project was suspended	No; magnitude and permanence of impacts not assessed (Element 2 4)

Table G-1. Continued

ID	HIA; Year; Location	Scope/Summary	Source of Evidence	Impacts/ Endpoints	Pathway of Impact	Characterization of Impact	Decision-making Outcome	Effectiveness of HIA	Minimum Elements of HIA Met? If no, what's missing
81	*Health Impact Assessment of the Port of Oakland*, 2010; Oakland, California	Evaluate the cumulative impacts of on-going Port of Oakland growth on the health of residents in West Oakland through multiple inter-related pathways	Literature review, community consultation, special collection (interviews, surveys, demographic analysis, modeling, GIS)	Health, environmental, economic, services, infrastructure	Air quality, physical activity, community/house-hold economics, healthcare insurance, land use, mobility/access to services, noise pollution, parks and recreation, safety and security, employment, social capital	Direction of impacts (positive health impacts, negative health impacts); Likelihood of impacts (definite impacts, probable impacts, undetermined impacts); Magnitude of impacts (percent population affected by noise pollution); Permanence of impacts (estimated traffic-related mortality rates); Distribution/equity of impacts (disparate health conditions of vulnerable communities; disproportionate effect of diesel particulate matter on vulnerable populations); Quantification of impacts (cumulative contributions of port and area traffic emissions using a roadway dispersion model; attributable mortality rate and adjusted mortality risk based on social position; modeled traffic noise levels/contours; percent population at risk of annoyance, sleep disturbance, and cognitive impairment, and deaths due to myocardial infarction); Qualitative analysis of impacts (pedestrian environment quality/ walkability, photo document-ation, assessment of housing values as a proxy for retail viability)	The Port of Oakland plays an important role in the movement of goods in the United States and can play both a positive and negative role in the health of West Oakland Multiple recommendations were made in each of the five areas examined (air quality, noise, transportation, retail, and labor) to mitigate negative health impacts, and seven of those were identified as more cross-cutting mitigation recommendations: considering health impacts in future planning, roadway improvements, and monitoring; exploring benefits of including noise emissions reductions with truck retrofits; creating a commercial corridor that meets community needs for healthy retail services; considering air pollution and noise mitigation for sensitive land uses; reducing the unemployment rate; diverting increasing tax revenue to specific community services; and improving Port operations	Undetermined	No; Elements 1 (informs a decision-making process), 2 1 (scoping), and 4 (monitoring plan) are missing; funding sources and sponsors were not identified (Element 5)

Table G-2. Summary Table of Select Data from HIAs in the Housing/Buildings/Infrastructure Sector

ID	HIA; Year; Location	Scope/Summary	Source of Evidence	Impacts/ Endpoints	Pathway of Impact	Characterization of Impact	Decision-making Outcome	Effectiveness of HIA	Minimum Elements of HIA Met? If no, what's missing
2	*Affordable Housing and Child Health: A Child Health Impact Assessment of the Massachusetts Rental Voucher Program*; 2005; Massachusetts	Evaluate the implications of the Massachusetts Rental Voucher Program (MRVP), a housing assistance and homelessness prevention program, and proposed MRVP changes for FY2006, for children's health and well-being	Literature review, special collection (interviews with stakeholders, local housing authorities survey, demographics analysis)	Health, behavioral, other (educational attainment)	Housing, community/ household economics, mobility/access to services, exposure to hazards	Direction of impacts (negative health impacts, positive health impacts, health impacts unclear); Magnitude of impacts; Quantification of impacts (percent increase in food insecurity, cost implications of educational impacts)	The majority of proposed changes would lead to budget trade-offs, disenrollment, and housing instability, all of which have adverse health effects. Proposals that lead to increased homelessness and housing instability would also result in increased education costs. Proposals that decrease tenant rent share decrease the need for budget trade-offs and children in families who cannot use their mobile vouchers to move out of high poverty areas may still experience the health benefits of increased household resources available for other basic needs	Direct effectiveness – provided testimony at legislative hearing; evidence provided was crucial to state's decision to not move forward	No; Elements 3 (recommendations and mitigations) and 4 (monitoring plan) missing; and documentation of sponsors, funding, and interviewed stakeholders not transparent (Element 5)
8	*Health Impact Assessment: South Lincoln Homes, Denver CO*; 2009; Denver, Colorado	Examine the redevelopment master plan for the Denver Housing Authority's South Lincoln Homes community in Downtown Denver for potential impacts the redevelopment may have on the health and wellbeing of the South Lincoln neighborhood	Literature review, community consultation, special collection (interviews/focus groups, survey, health survey, food audit, retail food availability survey, walk-ability audit, HDMT, demographics analysis, GIS)	Health, environmental, behavioral, economic, infrastructure, services, demographic	Social capital, mental health, cultural identity and equity, physical activity, land use, nutrition, parks and recreation, mobility /access to services, healthcare access, safety and security, air quality, noise pollution, house-hold/community economics, water quality, exposure to hazards, socio-economic status	Direction of impacts (positive health impacts); Distribution/ equity of impacts (impacts on neighborhood of low socio-economic status); Quantification of impacts (benchmarks of performance; walkability quantified using the Pedestrian Environmental Quality Index; resident health and availability of healthy food quantified via neighborhood surveys)	Concluded that the master plan included many sustainable design concepts that focus on health and well-being of the residents. Provided a very detailed series of recommendations to optimize positive impacts and mitigate negative impacts in five categories – social and mental wellbeing, natural environment, built environment and transportation, access, and safety. The inclusion of recommendations in the master plan document and/or further actions to be taken were noted. Some recommendations included improved health care access, walkability, pedestrian and traffic safety, infrastructure that promotes bicycling, social capital, and access to healthy foods; reduced parking demand and footprint; increased park and recreation spaces; and improved environmental conditions	Direct effectiveness – Mithun, one of the organizations involved in the HIA, is also the firm hired by the Denver Housing Authority to complete the master plan; the HIA was included in the Final Redevelopment Master Plan and HIA-related changes are incorporated in the master plan	No; permanence, magnitude and likelihood of impacts not assessed (Element 2 4); and documentation of funding sources in not transparent (Element 5)
12	*Community Health Assessment: Bernal Heights Preschool - An Application of the Healthy Development Measurement Tool (HDMT)*; 2008; San Francisco, California	Inform decision making processes related to the choice among three potential future locations of the Bernal Heights Preschool	Literature review, special collection (modeling, HDMT demographics analysis, GIS)	Health, behavioral, infrastructure, services, demographic	Childcare, housing, exposure to hazards, education, parks and recreation, physical activity, mobility /access to services, air quality, noise pollution, demo-graphics, social capital, nutrition	Direction of impacts (positive health impacts, negative health impacts); Distribution/equity of impacts (childcare for lower income families/retain families)	The majority of data available in the HDMT was not geographically specific enough to differentiate between the three potential locations under consideration, but the key findings point to a strong need for childcare, investment in schools and parks, valuing of community and social interactions, and retention of racially and socioeconomically diverse communities in the Bernal Heights neighborhood. No recommendations on preferred location were provided, but HIA suggested that Option C was the only location that allows the preschool to expand the number of children served and keep the preschool near Cortland Avenue	Undetermined – the Bernal Heights Preschool has been re-located out of the library, but it is not evident whether this is a temporary or permanent move	No; Element 4 (monitoring plan) missing; magnitude, likelihood, and permanence of impacts not assessed (Element 2 4); and source of funding not transparent (Element 5)

Table G-2. Continued

ID	HIA; Year; Location	Scope/Summary	Source of Evidence	Impacts/ Endpoints	Pathway of Impact	Characterization of Impact	Decision-making Outcome	Effectiveness of HIA	Minimum Elements of HIA Met? If no, what's missing
23	Health Impact Assessment of Modifications to the Trenton Farmer's Market (Trenton, New Jersey); 2007; Trenton, New Jersey	Examine several proposed changes to a farmers market in Trenton, New Jersey, including two being considered by the market's executive board (i.e., minor cosmetic changes and a major re-model) and a third suggestion (i.e., a market outreach strategy) and their impacts on patrons' nutrition and physical activity patterns, as well as the potential economic and social capital benefits for vendors and the surrounding community	Literature review, special collection (interviews, surveys, public meetings, demographics analysis, modeling)	Health, economic, services, behavioral	Nutrition, community economics, physical activity, social capital, preventative health services	Direction of impacts (impacts under each alternative were rated as "no change," "potentially beneficial," and "potentially harmful" for each pathway); Distribution/equity of impacts (based on major disparities in health status, risk factors, and food access, evaluated impacts to different population segments - within 2 miles of the market, city of Trenton, and Mercer County)	Market's executive board was only interested in making limited (cosmetic) changes, which would likely not significantly impact health; but the HIA offered recommendations to improve the alternative (such as equipping vendor stalls with electronic benefits transfer machines and ensuring the prepared food vendors offer healthy options) The HIA found that Alternative 2 (major remodel) could yield significant health impacts (economic and social capital), but would probably not improve consumption of fresh fruit and vegetables; Alternative 3 (outreach strategy) by comparison, had the best likelihood for improving nutrition Concluded a combination of the second and third alternatives posed the greatest potential benefit One overall recommendation – that the market offer nutrition education programs and market coupons to increase likelihood of selecting healthy options	Undetermined, but no effectiveness assumed – based on executive board's input on alternative scenarios, no effectiveness assumed	No; Element 4 (monitoring plan) missing; and magnitude, likelihood and permanence of impacts not assessed (Element 2 4)
28	The Rental Assistance Demonstration Project - A Health Impact Assessment 2012; United States	Examine the impacts of a proposed federal housing policy designed to address some of the systemic funding issues related to public housing on a number of health determinants that remained unanswered in legislative debates; and ensure that the evaluation of this pilot project comprehensively considered the health impacts of public housing-related policy decisions	Literature review, policy review, special collection (focus groups, surveys, demographics analysis)	Health, infrastructure, behavioral, economic	Housing, physical activity, community/household economics, mobility/ access to goods and services, nutrition, social capital, safety and security, exposure to hazards	Direction of impacts (negative health impacts, positive health impacts, mixed result impact); Magnitude of impacts (no people living in housing units if the pilot project is implemented more widely; could also impact the lives of individuals living on the edge of economic insecurity); Magnitude of specific impacts (negligible, minor, moderate, or major); Severity/ Permanence of impacts (high, moderate, or low); Likelihood of impacts (probable impacts, speculative impacts); Strength of evidence (plausible but insufficient evidence; likely but more evidence needed; causal relationship certain); Distribution/equity of impacts (impacts hard-to-house populations)	Found that RAD, as currently written, would have significant impacts on the health of public housing residents and communities, and the impacts were more negative than positive – especially if recommendations proposed in the HIA were not adopted Seven overarching recommendations were provided (e g, funding to improve existing public housing stock; keeping public housing "public;" preservation of public housing stock; funding for services and support for hard to house; a conversion oversight committee; including local resident association participation in review and decision making; an assessment, monitoring, and evaluation program to track implementation and effects of RAD); and 35 specific recommendations, such as requiring environmentally sustainable rehabilitation, expanding due process protections for public housing residents, requiring just cause evictions of residents, and limiting how far residents are relocated	General effectiveness – 2011 bill passed before HIA released, but HIA elevated health in a discussion that did not typically include health; impact of the HIA is on-going since RAD is a pilot project (i e, HIA will be used through end of the pilot to evaluate and monitor RAD's effects)	Yes

Table G-2. Continued

ID	HIA; Year; Location	Scope/Summary	Source of Evidence	Impacts/ Endpoints	Pathway of Impact	Characterization of Impact	Decision-making Outcome	Effectiveness of HIA	Minimum Elements of HIA Met? If no, what's missing
30	*Jack London Gateway Rapid Health Impact Assessment: A Case Study*; 2007; West Oakland, California	Examine a planned retail expansion and low-income senior housing development and address community concerns about air quality, noise, safety, and retail planning	Literature review, community consultation, special collection (demographics analysis)	Health, infrastructure, services, economic, demographic	Housing, air quality, noise, safety and security, social capital, nutrition, parks and recreation, physical activity, mobility/ access to services, public health services, social equity, household economics/liveli-hood, water quality, education, democratic process (participation in public decision making)	Direction of impacts (negative health impacts, positive health impacts); Distribution/equity of impacts (senior citizens impacted)	Found that, without mitigations, the project could lead to health concerns and made recommendations to mitigate potential negative impacts (e.g., air quality monitoring, installation of ventilation systems with modest filtration, a noise study, noise buffering, obtaining crime statistics, increasing private security talking with Neighborhood Crime Prevention Council, improving walkability, implementing traffic calming measures, creating a retail plan to meet community needs/wishes)	Direct effectiveness – when developer did not guarantee implementation of mitigations, two members of HIA Working Group testified before the Design Review Committee; as a result, a central ventilation system/air filters was installed, and building design was modified to orient the entry way through a noise-buffered courtyard; developer in discussion with the Neighborhood Crime Prevention Council and conducted a small survey to evaluate interest in retail usage HIA also sparked additional HIA work in area	No; magnitude, likelihood, and permanence (Element 2.4) not directly judged; studies suggested to further assess/quantify impact
35	*A Health Impact Assessment of Accessory Dwelling Unit Policies in Rural Benton County, Oregon*; 2011; Benton County, Oregon	Examine the impacts of five accessory dwelling unit policy options, ranging from restricting currently permitted uses to allowing construction of a complete accessory unit	Literature review, community consultation, policy review, special collection (focus groups, interviews, HDMT, GIS)	Health, infrastructure, services, other	Housing, mobility/ access to services, land use, social capital, secondary pathways: house-hold economics, healthcare access, lifestyle, mental health, physical activity, safety	Direction of impacts (positive health impacts, negative health impacts); Direction (positive health impacts, negative health impacts) and Magnitude (low to moderate, significant) of impacts and best/worst policy option by assessment area (housing, access to goods and services, social and family cohesion, transportation and mobility); Distribution/equity of impacts (population most benefited; affordable housing in communities that often suffer from higher rates of poverty); Quantification of impacts (projected ADUs annually)	Found that two of the five policy options would have positive health impacts – Option 3 (allowing smaller, more restrictive ADUs) due to the family-friendly living arrangements that ADUs encourage and Option 2 (restricting currently permitted uses such as medical hardship trailers and satellite bedrooms) due to restrictions on the number of dwelling units in rural parts of the County with poor accessibility to goods, services, and public transit options Despite the two policies' equal impacts, Option 3 was the more socially and politically preferred policy option, as it clearly benefited several vulnerable populations (disabled and elderly) and was in line with the desires of policymakers and public stakeholders to provide an avenue for family care Recommended that the Benton County Community Development Department consider the adoption of an ADU policy similar to Option 3 and provided recommendations to ensure the positive health benefits were optimized	Direct effectiveness – as a result of this HIA, Benton County's code was amended to allow ADUs	No; likelihood and permanence of impacts not assessed (Element 2.4)

G-14

Table G-2. Continued

ID	HIA; Year; Location	Scope/Summary	Source of Evidence	Impacts/ Endpoints	Pathway of Impact	Characterization of Impact	Decision-making Outcome	Effectiveness of HIA	Minimum Elements of HIA Met? If no, what's missing
36	*The Health Impact Assessment (HIA) of the Commonwealth Edison (ComEd) Advanced Metering Infrastructure (AMI) Deployment;* 2012; Northern Illinois (Commonwealth Edison Utility Territory)	Identify the impact of advanced metering infrastructure (AMI) deployment on the health of residential customers in the Commonwealth Edison (ComEd) service territory in Illinois, particularly vulnerable customers – the very young (birth to age 5), older individuals (age 65+), individuals with functional disability status including those with temperature sensitive conditions, individuals who are socially isolated, and individuals with limited English proficiency or literacy	Literature review, community consultation, policy review, special collection (surveys)	Health, environmental, behavioral	Air quality, community/house-hold economics, housing, exposure to hazards, health-care access/ insurance, nutrition/ food security, safety and security	Direction of impacts (negative health impacts, impact predicted but none seen in pilot or insufficient evidence available); Magnitude of impacts/size of at risk groups (all households with AMI, % of households); Severity/Permanence and Likelihood of impacts (moderate impact on few, moderate impact on medium number or strong impact on few, strong impact on medium number or moderate impact on many, strong impact on many); Distribution/equity of impacts (disproportionate impacts on vulnerable populations); Quality of evidence	Found that AMI implementation could result in several negative impacts including higher residential energy costs for vulnerable populations, economic incentives for customers to use less electricity when it is most needed for central air conditioning (i.e., critical peak pricing), and expedited disconnections and reconnections for nonpayment, as well as remote disconnections. The HIA provided five recommendations to help mitigate these negative impacts, including analysis of likely impacts on health and safety for clearly defined groups and at-risk residential customers, linking benefits and costs of proposed cost recovery on vulnerable customers, incentives for vulnerable households to optimize their use of electricity in time-based pricing programs, deployment of remote connection/disconnection functionality in a way that promotes health and safety of vulnerable customers, and robust consumer education and outreach	Direct effectiveness – after the HIA team provided testimony at a regulatory hearing, the Illinois Commerce Commission supported funding a robust consumer education system, maintained the current system requiring a site visit for disconnection for non payment, and determined that metrics designed to measure the impact of the technology on vulnerable populations be developed	Yes

G-15

Table G-2. Continued

ID	HIA; Year; Location	Scope/Summary	Source of Evidence	Impacts/ Endpoints	Pathway of Impact	Characterization of Impact	Decision-making Outcome	Effectiveness of HIA	Minimum Elements of HIA Met? If no, what's missing
40	*Unhealthy Consequences: Energy Costs and Child Health - A Child Health Impact Assessment of Energy Costs and the Low Income Home Energy Assistance Program*; 2007; Boston, Massachusetts	Evaluate impacts of both home heating and total home energy (including electricity, water heating, and cooking) costs and a federally-funded energy assistance program –the Low Income Home Energy Assistance Program (LIHEAP) –on the health of children	Literature review, policy review, special collection (interviews, demographics analysis)	Health (physical, mental, developmental), behavioral, economic, services	Community/ house-hold economics, exposure to hazards, education, housing, healthcare access/insurance, air quality, infect-ious disease, life-style, mental health, nutrition, safety and security	Direction of impacts (negative health impacts); Distribution/ equity of impacts (low-income families have disproportionate impacts; LIHEAP vulnerable households); Magnitude of impacts (number low-income households with children in Massachusetts who are likely LIHEAP eligible; number children in low-income families, who live below the poverty line, and are five years old and younger)	Current LIHEAP benefits, targeted to especially vulnerable populations, are helpful, but not sufficient in buffering families from the impact of high energy costs (although their situation would be even more precarious without this assistance) Six recommendations were developed that offer funding, programmatic, and data collection strategies to avoid the public health impact of unaffordable energy: 1 Federal government should fully fund LIHEAP at the maximum authorized level to allow an increase in both participation and benefit level; 2 Massachusetts state government should allocate supplementary funds for LIHEAP to increase benefit levels for vulnerable Massachusetts families; 3 Extend an initiative to clinicians and health care settings; 4 Consider an initiative to provide energy and utility assistance, through LIHEAP or other energy assistance programs, to eligible low-income families more quickly; 5 Enforce the existing requirement that utility commissions collect and report data on arrearages and utility disconnections to the Department of Telecommunications and Energy; 6 Energy assistance programs should explore the utility of a home energy insecurity scale to assess initial and subsequent energy self-sufficiency of households before and after receipt of energy benefits, providing a useful evaluation of the impact of these benefits	Direct effectiveness – members of the Working Group presented their findings to the state legislature in testimony before the joint committee on housing; HIA ultimately contributed to a decision to increase the level of funding to the program Groups in Rhode Island used the HIA report to advocate for increased levels of funding in that state, as well	No; Element 4 (monitoring plan) missing; permanence and likelihood of impacts not considered (Element 2 4); and sources of funding anonymous (Element 5)

Table G-2. Continued

ID	HIA; Year; Location	Scope/Summary	Source of Evidence	Impacts/ Endpoints	Pathway of Impact	Characterization of Impact	Decision-making Outcome	Effectiveness of HIA	Minimum Elements of HIA Met? If no, what's missing
48	*A Rapid Health Impact Assessment of the City of Los Angeles' Proposed University of Southern California Specific Plan*; 2012; Los Angeles, California	Examine how the proposed University of Southern California (USC) Specific Plan would impact measures of housing gentrification, and displacement and lead to changes in health for the communities around the USC campus, particularly low-income and vulnerable populations	Literature review, community consultation, policy review, special collection (residents panel, GIS)	Health, economic, infrastructure, demographic	Air quality, community/ house-hold economics, education, exposure to hazards, health-care access/ insurance, housing, infectious disease, mental health, mobility/access to services, nutrition, social capital	Direction of impacts (positive health impacts, negative health impacts); Likelihood of impacts (definite impacts, likely impacts); Distribution/ equity of impacts (people living close to USC impacted disproportion-ately, vulnerable populations most impacted by displacement /lack of affordable housing and increased housing costs without increase in wages); Quantification of impact (indices of gentrification and displacement)	If the USC Plan was implemented without mitigations, the HIA found that the result would be a high risk of gentrification, low vacancy rates, increased housing costs in the communities that surround the University, and further displacement of current low-income residents. Twelve recommendations were developed to mitigate the identified impacts, including developing an Affordable Housing Trust Fund, financing the preservation of currently affordable units whose covenants will expire in the next five to twenty years, improving the local hiring policies, paying a living wage and hiring local, nonstudent residents for these jobs	Direct effectiveness – HIA report was submitted to the Planning Commission and Los Angeles City Council to consider in the USC Specific Plan decision-making process; USC agreed to invest $20 million in affordable housing, use local and disadvantaged hiring, increase the number of net new student beds on campus, and provide legal assistance, and business assistance	No; Element 4 (monitoring plan) missing; magnitude and permanence of impacts not assessed (Element 2 4)
50	*Anticipated Effects of Residential Displacement on Health: Results from Qualitative Research*; 2005; San Francisco, California	Examine the Trinity Plaza Redevelopment, which proposed to demolish an older apartment building with over 360 rent-controlled units and replace it with 1,400 market-rate condominiums and the potential effects of eviction on health and well-being of tenants	Literature review, community consultation, policy review, special collection (focus groups)	Health, economic	Community/ house-hold economics, exposure to hazards, healthcare access/insurance, housing, infectious disease, mental health, nutrition, social capital	Direction of impacts (negative health impacts); Distribution/ equity of impacts (vulnerable populations and populations more impacted identified)	Officials from the Department of City Planning initially concluded that redevelopment of the site would not have adverse housing impacts, because the proposal increased the total number of dwelling units; however, public testimony from residents/tenant advocates and the results of this study show otherwise. The health impacts of displacement are real and provide compelling evidence for preventing the loss of existing affordable housing due to redevelopment through rent subsidies, targeted maintenance subsidies to landlords who supply affordable housing [to prevent demolition]; and general maintenance subsidy to all landlords to make housing more affordable	Direct effectiveness – Department of City Planning revised their determination for the Trinity Plaza proposal and required the environmental impact report to analyze residential displacement and indirect impacts on health; the developer implemented an alternative that allowed the current residents to remain and provided 360 permanently rent-controlled units	No; Elements 2 3 (health baseline) and 4 (monitoring plan) missing; magnitude, likelihood and permanence of impacts not addressed (Element 2 4); conclusions are not transparent & recommendation does not address problem facing the study (rather it addresses the larger problem of loss of affordable housing; Element 2 5); and documentation of funding sources not transparent (Element 5)

Table G-2. Continued

ID	HIA; Year; Location	Scope/Summary	Source of Evidence	Impacts/ Endpoints	Pathway of Impact	Characterization of Impact	Decision-making Outcome	Effectiveness of HIA	Minimum Elements of HIA Met? If no, what's missing
52	*29th St. / San Pedro St. Area Health Impact Assessment*; 2009 Los Angeles, California	Ensure that health impacts were considered in the development plan for The Crossings at 29th Street - an 116-acre development that included affordable housing and retail and community space - and in the broader policies impacting redevelopment in the area	Literature review, community consultation, policy review, special collection (survey, pedestrian environment and public transit field assessment, GIS)	Health, infrastructure, services, economics, behavioral	Education, healthcare access/ insurance, housing, land use, mobility /access to services, noise pollution, nutrition, physical activity, safety and security, parks and recreation social capital	Direction of impacts (positive health impacts, negative health impacts); Likelihood of impacts (definite impacts, probable impacts, speculative impacts); Distribution/equity of impacts (socioeconomic status of population in study area; population over-represented in pedestrian deaths); Quantification of impacts (pedestrian quality assessment)	Found the development had the potential to bring many benefits to the health and well being of local residents in the area; however, the HIA offered a number of recommendations to maximize the positive health impacts and minimize the negative impacts of the development in the following areas - housing; pedestrian safety, neighborhood walkability, and public transit; health services and food retail; education; and parks and recreation. Also recommended that air quality, noise, and chemical contamination of groundwater and soil from industrial sources (not included in the HIA) and their health impacts should be studied in depth as part of the EIR process	Direct effectiveness - the City Council representative in the area agreed to implement some of the HIA recommendations and the developer agreed to reduce the cost of housing in future phases of the development per the HIA findings; all 450 housing units in the first phase of development were categorized as affordable	No; Element 4 (monitoring plan) missing; and permanence and magnitude of impacts not assessed (Element 2.4)
57	*Lowry Corridor Phase 2 Health Impact Assessment*; 2007; Minneapolis, Minnesota	Analyze the potential health effects of Phase 2 development of the Lowry Avenue Corridor Project, a five-mile thoroughfare located north of downtown Minneapolis	Literature review, policy review, other (demographics analysis, GIS)	Health, environmental/ ecosystem, economic, infrastructure, other (social)	Community/house-hold economics, mobility/access to services, housing, lifestyle, physical activity, mental health, nutrition, safety and security, social capital, water quality	Direction of impacts (positive health impacts, negative health impacts); Severity (permanence) of impacts (low, medium or high); Likelihood of impacts (probable impacts, possible impacts, speculative impacts); Distribution/equity of impacts (socioeconomic status of population in area; yes or no to whether each of the specific health determinants would have differential impacts on groups)	Overall, the impacts of the Lowry Corridor Phase 2 construction and redevelopment project were determined to be positive, with exception of potentially negative impacts from construction-related access to businesses and property acquisition. Recommendations were identified for each determinant of health, along with the coordinating entity(ies) responsible for implementing the recommendation, where applicable. The 29 recommendations centered around four primary themes - social connections, right-of-way design, stabilizing the neighborhood, and physical activity	Direct effectiveness - raised the project manager's awareness of project health impacts, which led to successful application for funding through the Non Motorized Transportation Pilot Program to purchase and place countdown timers at key intersections, bike racks at key public buildings, and markers to encourage pedestrian traffic; recommended incorporating HIAs into policymaking and planning for infrastructure in Hennepin County	No; Element 2.2 (stakeholder involvement) missing; unclear whether best available evidence is used to judge potential health impacts and make conclusions and recommendations (Elements 2.4 & 2.5)

Table G-2. Continued

ID	HIA: Year; Location	Scope/Summary	Source of Evidence	Impacts/ Endpoints	Pathway of Impact	Characterization of Impact	Decision-making Outcome	Effectiveness of HIA	Minimum Elements of HIA Met? If no, what's missing
61	Hospitals and Community Health HIA: A Study of Localized Health Impacts of Hospitals; 2008; Atlanta, Georgia	Built upon the Atlanta BeltLine HIA to retrospectively examine the localized health impacts of Piedmont Hospital - one of the major anchor institutions along the Peachtree Corridor in Atlanta, Georgia - and prospectively examine how plans for future growth could change those impacts	Literature review, community consultation, policy review, special collection (surveys, walkability audit, demographic analysis, GIS)	Health, environmental, behavioral, economic, infrastructure, services, demographic	Air quality, community/ household economics, healthcare access, mobility/access to services, noise pollution, nutrition, parks and recreation, physical activity, public health services, safety and security, social capital	Direction of impacts (negative health impacts, positive health impacts); Distribution of impact (vulnerable populations identified and block groups given a vulnerability rating; two most vulnerable block groups directly adjacent to the hospital may be disproportionately affected by the negative health impacts); Quantification of impacts (vulnerability score of block groups calculated using 6 indicators)	Taking into account the BeltLine redevelopment corridor improvements and streetcar projects, the HIA identified a number of general recommendations to increase opportunities for health and mitigate negative health impacts, such as level sidewalks with ample buffers between pedestrians and traffic, improved lighting, bike lanes with sufficient room for bicyclists/cars, consideration of pedestrian and bicycle access when making future decisions regarding hospital planning, improved communication between the hospital and community groups, increased transit usage, improved signage, use of universal design methods to develop safe connections, improved intersection safety, and improvements to the pedestrian environment	General effectiveness – findings and recommendations were presented to neighborhood organizations, city officials, and county commissioners; residents took ownership and formed a working group that reached out to Piedmont Hospital requesting better community relations and implementation of the HIA recommendations; hospital staff responded positively and are meeting regularly with the group (Source: RWJF)	No; Element 4 (monitoring plan) missing; and magnitude, likelihood, and permanence of impacts not judged (Element 2.4)
70	Pathways to Community Health: Evaluating the Healthfulness of Affordable Housing Opportunity Sites Along the San Pablo Avenue Corridor Using Health Impact Assessment; 2009; Oakland, California	Assess the health impacts associated with the San Pablo Avenue Specific Plan for three sites proposed to be included in a campaign for affordable housing and encourage the healthfulness of the San Pablo Area Specific Plan and eventual site development	Literature review, community consultation, special collection (focus groups, modeling, food retail and pedestrian environment evaluations, HDMT, GIS)	Health, environmental, economic, infrastructure, services	Air quality, noise pollution, exposure to hazards, housing, mobility/access to services, education, nutrition, parks and recreation, physical activity, safety and security, community economics, social capital	Direction of impacts (positive health impacts, negative health impacts); Comparison of impacts by site; Distribution/ equity of impacts (mapping of proximity measures; populations sensitive to environmental noise/air pollution, and access to community/senior centers; populations more at risk for traffic collisions); Quantification of impacts (Retail Food Environment Index score; PM2.5 concentrations attributable to roadway traffic; Pedestrian Environment Quality Index scores; noise modeling; mapping of noise contours); Permanence and Magnitude of air quality impact (pre-mature mortality due to PM2.5 exposure per million population)	Found that the health impacts at the three assessed affordable housing sites would be similar overall. Slight differences in impacts were noted for concentrated poverty, violence and crime, and access to a full service supermarket/community center. A more significant difference between the sites was determined for quality of schools (i.e., the Albertsons site lacks access to high-quality public schools) Provided a number of recommendations designed to mitigate the expected adverse health impacts, such as increasing space for healthy retail and public services, offering reduced-cost transit passes to residents, incentives for car-sharing, performing a needs assessment of local park programming, filling gaps in park access, improving park maintenance and security, build a new public elementary school to handle the population growth associated with the plan, incorporating bike lanes and traffic calming features, implementing strategies for reducing air and noise impacts, and implementing mixed-use development	Undetermined, but general effectiveness assumed – a letter to City Council and city staff with health-based recommendations was considered during revisions to the Draft San Pablo Avenue Specific Plan to incorporate all public comments to date	No; Element 4 (monitoring plan) missing; likelihood of impacts not assessed and permanence and magnitude only assessed for air quality impacts (Element 2.4)

Table G-2. Continued

ID	HIA; Year; Location	Scope/Summary	Source of Evidence	Impacts/ Endpoints	Pathway of Impact	Characterization of Impact	Decision-making Outcome	Effectiveness of HIA	Minimum Elements of HIA Met? If no, what's missing
76	*A Rapid Health Impact Assessment of the Long Beach Downtown Plan* 2011; Long Beach, California	Ensure decisions in the City of Long Beach Downtown Plan and Long Beach Downtown Plan Environmental Impact Report account for impacts to low-income and vulnerable populations in the areas of housing and employment	Literature review, policy review, special collection (demographics analysis, GIS)	Health	Housing, community/ household economics, exposure to hazards, noise pollution, mental health	Direction of impacts (negative health impacts); Likelihood of impacts (definite impacts, probable impacts); Distribution/ equity of impacts (lower SES and minority populations disproportionately impacted)	Found that the Long Beach Downtown Plan the Downtown Plan did not identify housing or employment mitigation measures to offset the Plan's significant health impacts, nor did it accommodate the needs of Long Beach's most vulnerable residents. Recommended adoption of the Affordable Housing Community Benefits (addition of 511 very low income apartments and 375 moderate income condominiums) and Local Hiring Community Benefits and Project Labor Agreements (hiring preference and requirements for local lower and moderate income residents leading to increases in income, improved job autonomy and reduced unemployment and poverty) proposed by the Long Beach Downtown Plan Community Benefits Analysis	None – HIA findings were used to advocate for changes in the proposed Downtown Plan, but the Long Beach City Council approved the Plan without taking into account the findings and recommendations in the HIA	No; Element 4 (monitoring plan) is missing
80	*HOPE VI to HOPE SF San Francisco Public Housing Redevelopment: A Health Impact Assessment*; 2009; San Francisco, California	Explore the positive and negative health impacts of past Housing Opportunities for People Everywhere (HOPE) VI redevelopment at two sites –Bernal Dwellings and North Beach Place - with the aim of finding opportunities to address existing problems and informing future public housing redevelopment in the HOPE SF Program	Literature review, policy review, community consultation, special collection (interviews, surveys, demographics analysis, modeling, GIS)	Health, behavioral, infrastructure, services	Housing, mobility/ access to services, safety and security, physical activity, social capital, public participation, exposure to hazards, nutrition, air quality	Direction of impacts (positive health impacts, negative health impacts); Likelihood of impacts (definite impacts); Distribution/ equity of impacts (socio-economic status of public housing population; population likely to be impacted disproportionately from displacement; population sensitive to availability of affordable housing); Quantification of impacts (modeled traffic noise levels and particulate matter 2.5 emissions and associated health impacts)	Provided a long list of recommendations to improve health at the HOPE VI sites, as well as additional recommendations for on-going HOPE SF redevelopment. HOPE VI recommendations were provided in the following areas: healthy housing and environmental health, displacement, social cohesion, crime and safety, youth programs and services, and healthy eating and active living. Select HOPE SF recommendations included broad stakeholder participation in discussions of how to improve health and address existing health disparities; design elements to improve safety; use high-quality healthy building materials; redevelop in stages to minimize the disruption associated with relocation; ensure adequate space, programming, and access to healthy foods and opportunities for active living; be mindful of the diversity of the residents; and listen to the residents	General effectiveness at a minimum, but direct effectiveness assumed – HIA was included in the City and County of San Francisco's discussions about HOPE SF; particularly HIA results were used in discussions around social cohesion, displacement, programs and services, and crime in the HOPE SF process	No; magnitude and permanence of impacts not assessed (Element 2.4)

G-20

Table G-3. Summary Table of Select Data from HIAs in the Land Use Sector

ID	HIA; Year; Location	Scope/Summary	Source of Evidence	Impacts/ Endpoints	Pathway of Impact	Characterization of Impact	Decision-making Outcome	Effectiveness of HIA	Minimum Elements of HIA Met? If no, what's missing
3	*Health Impact Assessment - Derby Redevelopment, Historic Commerce City, Colorado*, 2007; Historic Commerce City, Colorado	Evaluate potential impact of Derby's redevelopment on physical activity and nutrition behaviors of the population of historic Commerce City	Literature review, community consultation, policy review, special collection (consultants, walkability study, traffic assessment, PhotoVoice, community forums, survey, demographics analysis, GIS)	Health, behavioral, infrastructure, demographic	Safety and security, physical activity and fitness, nutrition, land use, social capital	Direction of impacts (positive health impacts); Magnitude of impacts (variable, high); Likelihood of impacts (definite impacts, probable impacts, speculative impacts); Distribution/equity of impacts (impacts evaluated on residents of Derby and the surrounding Commerce City; universal design favorable for all demographic groups)	Supported redevelopment plans for Derby, Colorado and concluded that the Derby Sub-Area Master Plan, Planned Unit Development zoning ordinance, and Design Guidelines would create physical conditions in Derby that foster active living, and to a lesser extent, healthy eating. Recommended a phased implementation, preparing a bicycle and pedestrian master plan, integrating green space and open space into existing plans, establishing a Clean and Safe Initiative, promoting affordable housing with universal design features, upgrading transit service and transit facilities, and developing an implementation plan for elements of the redevelopment that are within city control	General effectiveness assumed at a minimum – the Health Department's full participation was invited in the redevelopment team	No; Element 4 (monitoring plan) missing
6	*Health Impact Assessment Report: Alcohol Environment - Village of Weston, WI*, 2011; Village of Weston, Wisconsin	Assess the impact of an alcohol policy on the community's health, specifically underage drinking and drinking and driving behaviors. While there was no specific policy under review at the onset of the project, the potential impacts of a retail outlet density policy, specifically a limit on future Class A alcohol licenses, on community health and development were assessed	Literature review, policy review, community consultation, special collection (stakeholder interviews, focus groups, modeling, community surveys, GIS, photomapping, demographics analysis)	Health, behavioral, economic, other	Land use (alcohol outlet density), lifestyle, safety and security, policy (alcohol access)	Impact of proposed alcohol outlet density policy implementation on alcohol consumption, underage drinking, drinking and driving; Direction, Permanence and Magnitude of impacts (positive, moderate impact on medium number of people); Likelihood of impacts (probable impacts); Distribution/equity of impacts (populations impacted more, uncertain); Quality of evidence (many strong studies, one or two good studies); Projected outcomes and impacts (positive or no effect) of policy implementation on indicators outlined in scoping phase	Data provided evidence that alcohol density not only impacts alcohol consumption, but also underage drinking and driving behaviors. Positive impacts of alcohol (social connection and revenue source) outweighed by the negative impacts (health, disease, injury, death, etc). Primary recommendations included: a moratorium on future Class A alcohol licenses, development of a Policy Exemption Committee, and development of an Alcohol License Review Board. Secondary recommendations included gathering consistent health related data among youth in the school district (i.e., CDC Youth Risk Survey), and gaining support of the Marathon County Board of Health for future HIA projects within the County	Undetermined – recommendations were presented to the Village Board, residents, alcohol prevention professionals in Wisconsin, and the Marathon County Board of Health, but no evidence of a policy change being considered by the Board to date; regardless of a decision to adopt the recommendations, the HIA was successful in building important relationships to further the discussion about alcohol misuse prevention in the community	Yes; although funding source not clearly identified

Table G-3. Continued

ID	HIA; Year; Location	Scope/Summary	Source of Evidence	Impacts/ Endpoints	Pathway of Impact	Characterization of Impact	Decision-making Outcome	Effectiveness of HIA	Minimum Elements of HIA Met? If no, what's missing
7	*Eastern Neighborhoods Community Health Impact Assessment Final Report*; 2007; San Francisco, California	Assess the health benefits and burdens of development, land use plans, and zoning controls in several San Francisco neighborhoods, including the Mission, South of Market, and Portero Hill	Literature review, community consultation, policy review, special collection (interviews, focus groups, surveys, HDMT, GIS)	Health, environmental/ ecosystem, behavioral, economic, infrastructure, services, demographic	Environmental stewardship, safety and security, mobility/ access to goods and services, social capital, housing, community/ household economics, infrastructure, land use, community participation, noise	N/A; HDMT tool developed out of the ENCHIA process can be used to quantify and prioritize impacts of development on health	A formal assessment of the development plans was not completed due to delays in the planning process; however, the ENCHIA process concluded with the creation of San Francisco's Healthy Development Measurement Tool (HDMT) for future plan and project evaluation – a set of metrics to assess the extent to which urban development projects, plans, and policies affect health	General effectiveness at a minimum, but direct effectiveness assumed – San Francisco Planning Dept committed to using the HDMT indicators and development criteria, where possible, in developing the Eastern Neighborhood rezoning and area plans; the HIA broadened participant understanding of how development affects health and created a practical tool (HDMT) for evaluating the health impacts associated with development	No; due to delay in development planning, Elements 2 4-4 (judgement of impacts, synthesis of evidence, recommendations, and monitoring plan) unable to be completed
10	*Health Impact Assessment: An Analysis of Potential Sites for a Regional Recreation Center to Serve North Aurora, Colorado*; 2010; North Aurora, Colorado	Inform a policy decision about the specific location of a regional recreation center in North Aurora, identify impacts to health, and provide recommendations for the Aurora Residents for Recreation Task Force (ARRTF), City Planners, and City Council	Literature review, special collection (survey, demographics analysis, GIS)	Health, behavioral, economic, services	Physical activity, mobility/access to services, safety and security, social capital, community economics	Direction of impacts (positive health impacts); Likelihood of impacts (definite impacts); Magnitude of impacts (portion of population that would benefit based on proximity); Distribution/equity of impacts (area deficient in recreational space and with a high percentage of demographic groups at risk for disease and getting too little physical activity)	Concluded that a new regional recreation facility in North Aurora will have a positive health impact on the community In light of the health disparities and potential cost to the City to provide equitable access to all North Aurora residents, the Fitzsimons Centerpiece site was recommended for the recreation facility (after the ideal location was determined to be unviable because owners were not willing to sell or lease the property) Also provided a list of recommended criteria to be incorporated regardless of which site is selected, as well as site-specific recommendations	Undetermined – the location recommended by the HIA was later determined not to be a viable option and therefore was not the chosen site for the recreational facility; it is not apparent, however, whether any of the site specific recommendations from the HIA were implemented in the design, many of which related to access (specifically for north Aurora residents)	No; Elements 2 2 (stakeholder input) and 4 (monitoring plan) missing; evidence of a scoping phase not apparent (Element 2 1); permanence of impacts not assessed (Element 2 4); and funding and roles not transparent (Element 5)

G-22

Table G–3. Continued

ID	HIA; Year; Location	Scope/Summary	Source of Evidence	Impacts/ Endpoints	Pathway of Impact	Characterization of Impact	Decision-making Outcome	Effectiveness of HIA	Minimum Elements of HIA Met? If no, what's missing
13	St. Louis Park Comprehensive Plan - Health Impact Assessment; 2011; St. Louis Park, Minnesota	Assess the St. Louis Park Comprehensive Plan to ensure that public health is considered within the plan	Policy review, special collection (GIS)	Health, environmental/ ecosystem, behavioral	Air quality, water quality, land use, parks and recreation, physical activity, nutrition, housing, mobility/ access to services, safety and security, exposure to hazard, social capital, healthcare access/ insurance, noise pollution	Direction of impacts (positive health impacts); Distribution/ equity of impacts (populations most vulnerable to lead toxicity); Quantitative analysis for established benchmarks (buffer analysis/proximity measures; total market value for housing compared to average median income level)	Overall, the Comprehensive Plan embraced public health throughout its respective chapters. The HIA provided recommendations (for incorporation into the City's Comprehensive Plan or other planning initiatives) to ensure health is addressed in planning efforts; they fall into three categories: Physical Activity and Access to Healthy; Personal Health and Safety; and Neighborhood/Community Health Recommendations touched on maintaining tree canopy/views of greenery, park and trail accessibility, pedestrian lighting, buffering major roads from sensitive uses, access to healthy food, housing affordability, protection from air/ water pollution, transit accessibility, and adopting complete streets policy	General effectiveness at a minimum, although direct effectiveness possible – the report was presented to the Planning Commission and adopted as a planning tool for consideration when updating the Comprehensive Plan	No; Elements 2.2 (stakeholder input) and 2.3 (baseline health conditions) missing (although stakeholder involvement not part of desk-based HIAs); magnitude, likelihood, and permanence of impacts not assessed (Element 2.4); and does not adequately acknowledge sources of data (Element 2.5)
15	Health Impact Assessment: Key Recommendations of the Northeast Area Plan; Unknown (possibly 2007); Columbus, Ohio	Evaluate the six key recommendations of the City of Columbus Northeast Area Plan with respect to physical activity for residents of the Northeast area	Literature review	Health, behavioral, environmental	Physical activity; secondary pathways: air quality, social capital, safety and security, mobility/ access to services, parks and recreation	Direction of impacts (positive health impacts; long term outcomes)	Recommended specific implementation strategies or features of each of the six key Area Plan recommendations that foster physical activity, including mixed-use planning and complete street tactics that capitalize on existing community centers, job centers, parks, and bike trails; and urban design components that emphasize pedestrian access and aesthetics	Undetermined – the HIA was the beginning of a working relationship between two City of Columbus departments (Public Health and Planning), and per the HIA, the relationship is just as important as the results of the HIA itself	No; Elements 2.2 (stakeholder involvement) and 4 (monitoring plan) missing; magnitude, permanence, likelihood, distribution of impacts not assessed (Element 2.4); baseline health conditions not addressed very well (Element 2.3); documentation of funding sources and HIA point-of-contact not transparent (Element 5)
16	Yellowstone County/City of Billings Growth Policy Health Impact Assessment; 2010; Yellowstone County, Montana	Take a retrospective look at the Growth Policy that was adopted in 2003 in order to identify ways to make health a part of the decision making process regarding community growth by predicting health consequences, informing decision makers and public about health impacts, and providing realistic recommendations to prevent/ mitigate negative health outcomes	Literature review, community consultation, policy review, special collection (focus groups, experts)	Health, services (emergency responders), infrastructure, economic	Physical activity, safety and security, social capital, nutrition, mobility/ access to services, housing, community/ household economics, land use	Direction of impacts (each Growth Policy strategy was evaluated as a positive, negative, or no effect health strategy); Distribution/equity of impacts (differences in access to affordable/nutritious foods and services by socioeconomic status [e.g., food deserts]; populations least likely to meet physical activity recommendations)	Identified key strengths and weaknesses of the Growth Policy as it pertains to health. Recommendations in several areas that could increase positive health outcomes and decrease or mitigate negative health outcomes - emergency preparedness, access to healthy foods, pedestrian and traffic safety, physical activity, social capital, safety and crime, affordable housing, and living wage jobs – and a recommendation made to incorporate a Community Health section in the revised 2008 Growth Policy	Direct effectiveness – recommendations were provided to the governing bodies of the Growth Policy for use during the revision process and ultimately all recommendations were implemented with minor changes; health outcome evaluation to be conducted in 2011	No; magnitude, likelihood and permanence of impacts not assessed (Element 2.4)

Table G-3. Continued

ID	HIA; Year; Location	Scope/Summary	Source of Evidence	Impacts/Endpoints	Pathway of Impact	Characterization of Impact	Decision-making Outcome	Effectiveness of HIA	Minimum Elements of HIA Met? If no, what's missing
18	*Knox County Health Department Community Garden Health Impact Assessment: Recommendations for Lonsdale, Inskip and Mascot*; Unknown; Knox County, Tennessee	Inform policy decisions related to the placement and maintenance of community gardens in Knox County, Tennessee and to objectively present the facts surrounding community gardens and why zoning code should be changed if needed in order to support their placement within residential and nonresidential communities	Literature review, policy review, special collection (demographics analysis)	Health, behavioral, economic	Land use, nutrition, mobility/access to services, social capital, community/household economics, safety and security	Direction of impacts (positive health impacts); Distribution/ equity of impacts (food availability, affordability, and justice for economically disadvantaged; all three pilot neighborhoods are of low socioeconomic status)	Community gardens offer many benefits (enhanced nutrition and physical activity), but there are challenges involved (zoning, siting, water access, security, and community interest) Lonsdale and Inskip were determined to be best suited for community gardens out of the three pilot neighborhoods because both are tight-knit communities with ample water sources and sidewalks that provide easy access to the garden site Going forward, the HIA recommended that areas of below average food access/availability and low SES be identified on a county-wide basis to prioritize selection of communities for community gardens General recommendations included: siting gardens in food deserts to increase food access and availability; siting gardens in low SES communities to increase food affordability and food justice; siting gardens near water access; and siting gardens near gardeners	Direct effectiveness – upon presentation of the HIA report, the zoning was changed to support the placement of gardens in designated areas of Knox County	No; Elements 2 2 (stakeholder involvement) and 4 (monitoring plan) missing: the presence of a scoping phase (Element 2 1) not evident; and magnitude, likelihood, and permanence of impacts not assessed (Element 2 4)
19	*Alaska Outer Continental Shelf - Beaufort Sea and Chukchi Sea Planning Areas, Oil and Gas Lease Sales 209, 212, 217, and 221 Draft Environmental Impact Statement; Appendix J - Public Health*; 2008; Alaska	Examine the health impacts of the proposals for oil and gas leasing in the Beaufort and Chukchi seas, as well as the 10 alternatives to these proposed actions addressed in the EIS	Literature review, policy review, community consultation, special collection (interviews, modeling)	Health, environmental/ ecosystem, behavioral, other (cultural), economic	Production activities; air quality; water quality; oil spill cleanup; habitat loss; seismic survey; community/ household economics; vessel and aircraft noise pollution; climate change; secondary impacts: social capital, safety and security, nutrition, exposure to hazards, infectious disease, lifestyle, physical activity, mobility/access to services; visuo-spacial changes	Direction of impacts (negative health impacts, positive health impacts); Distribution/equity of impacts (impacts on native Alaskans and vulnerabilities of population); Magnitude of impacts (negligible, minor, moderate, major); Likelihood of impacts (potential impacts, anticipated impacts, cumulative impacts); Permanence of impacts (timeline/extent of some impacts noted)	The No Action Alternatives would have no direct/indirect impacts to public health, but relative to the alternatives that involve oil and gas leasing, would offer the least revenue and employment Identified a number of health effects for the other alternatives, which involved oil and gas leasing to different extents Subsistence harvest disruptions were a major concern and could impact general health and wellbeing, diet and nutrition, injury rates, and rates of chronic diseases Identified standard mitigation measures assumed to be in place from existing government policies/programs, as well as potential new mitigation measures recommended to address newly identified health effects associated with the alternatives in the EIS New mitigation measures included public health baseline assessment and health monitoring; subsistence and nutrition monitoring/mitigation; air quality baseline assessment, modeling, monitoring, and mitigation; best practices to prevent OCS discharges; socioeconomic monitoring and mitigation; health impact evaluation for siting of on-shore infrastructure; and noise related monitoring and mitigation	General effectiveness assumed – one of the first examples of a US federal agency including information from HIA in an EIS; MMS agreed to consider mitigation measures in the HIA for this and other region-specific EIS; Arctic lease sales were cancelled in March 2010 after a ruling that the 2007-2012 national offshore oil and gas leasing program did not properly evaluate environmental sensitivity or benefits/risks of development; Beaufort and Chukchi leases possible in new leasing program, with areas important to the environment and/or subsistence conditions excluded	No; Element 4 (monitoring plan) missing; and documentation of funding sources not transparent (Element 5)

G-24

Table G-3. Continued

ID	HIA; Year; Location	Scope/Summary	Source of Evidence	Impacts/ Endpoints	Pathway of Impact	Characterization of Impact	Decision-making Outcome	Effectiveness of HIA	Minimum Elements of HIA Met? If no, what's missing
20	*Divine Mercy Development Health Impact Assessment*; 2011; Fairbault, Minnesota	Inform recommendations on incorporating health and climate change indicators into the Minnesota Environmental Assessment Worksheet (EAW) used in the environmental review process	Literature review, policy review, special collection (demographics analysis; GIS)	Health, environmental/ ecosystem	Air quality, land use, mobility/access to services, water quality, safety and security, housing, nutrition, noise pollution, parks and recreation, trees and vegetation, physical activity, social capital	Direction of impacts; Distribution/equity of impacts (populations vulnerable to certain climate change-related health impacts; park proximity and physical activity; benefits of public transit for disadvantaged populations; vulnerable populations to noise pollution)	Recommended actions to reduce or eliminate the health impacts of the Divine Mercy Development, such as monitoring the adjacent feedlot for air quality issues, higher residential/commercial densities, sufficient pedestrian infrastructure, housing for renters and those below median household income, recreational facilities and bike trails, 40% tree canopy, incorporating transit service, tracking accidents to identify problem intersections, incorporating grocery store and community garden features, and additional storm water management features. The HIA found health indicators in the categories of air quality, water, and noise were thoroughly analyzed in the EAW; those in the categories of housing, food, and safety were almost entirely absent; and the transportation, parks, land development, and trees and vegetation categories were discussed, but not thoroughly analyzed. General recommendations on incorporating public health and climate change indicators into the EAW were to be included in a separate report	Undetermined – HIA Report provided to Divine Mercy Development and separate report on incorporating health and climate change in the EAW process presented to MN Environmental Quality Board, but impact unknown	No; Elements 2.2 (stakeholder input) and 4 (monitoring plan) missing; and magnitude, likelihood, and permanence of impacts not assessed (Element 2.4)
21	*Fort McPherson Rapid Health Impact Assessment: Zoning for Health Benefit to Surrounding Communities During Interim Use*; 2010; Atlanta, Georgia	As part of a project to bring a Health in all Policies (HiAP) perspective into the baseline realignment and closure process for Fort McPherson and assessed the zoning provisions that govern permitted uses of land, green space, and transportation to gauge their effect on health	Literature review, community consultation), special collection (windshield tours, meetings, GIS)	Health	Land use, nutrition, physical activity, lifestyle, social capital	Direction of impacts (positive health impacts, negative health impacts); Likelihood of impacts (speculative impacts, possible impacts, probable impacts, definite impacts); Distribution/equity of impacts (health impacts of green space connectivity and accessibility on certain populations); Strength of evidence	Redevelopment may result in positive health impacts for the surrounding communities and eventually new residents, but interim use of the property could have a crucial impact on community health. Recommended some enhancements to promote positive impacts during this time, including permitting community gardens, small-scale farming, and farmers markets in green space and designated areas; permitting the use of existing facilities for meetings and programming; limiting fast food restaurants and establishments that serve or sell alcoholic beverages; limiting outdoor and storefront advertising; enforcing federal policy prohibiting tobacco advertising in the vicinity of schools, parks, and playgrounds; and prohibiting bars/restaurants that do not support the state tobacco policy	General effectiveness at a minimum, although direct effectiveness possible – the design contract has a requirement to consider Health in all Policies (HiAP) and the preliminary redevelopment design includes better connectivity and the addition of athletic fields and a grocery store; not evident whether this was a direct impact of the HIA or the greater HiAP effort in general	No; Elements 2.3 (baseline health conditions) and 4 (monitoring plan) missing; and magnitude and permanence of impacts not assessed (Element 2.4)

Table G-3. Continued

ID	HIA; Year; Location	Scope/Summary	Source of Evidence	Impacts/ Endpoints	Pathway of Impact	Characterization of Impact	Decision-making Outcome	Effectiveness of HIA	Minimum Elements of HIA Met? If no, what's missing
22	*Re: November 10th Merced County General Plan Update (MCGPU) Preferred Growth Alternative Decision*; 2009; Merced County, California	Examine the two growth alternatives being considered for the Merced County General Plan Update – one that focused development in existing urban areas and another that would allow for the creation of new towns in the county – and associated health impacts	Literature review, policy review	Health, environmental, economic	Mobility/access to retail and services, physical activity, social capital, safety and security, nutrition, land use, community/household economics, water quality/availability, climate change, air quality, exposure to hazards	Direction of impacts: Growth in existing urban areas (positive health impacts); Developing new towns (negative health impacts); Likelihood of impacts (definite impacts, probable impacts, speculative impacts)	Identified links between development decisions and health, existing conditions related to health in Merced County, and ways that the General Plan Update growth alternatives would potentially impact health outcomes for current and future county residents. Recommended focusing population growth and development in areas where there is existing urban development, infrastructure, and municipal services; promoting higher residential densities in urban areas; and implementing a Tax Revenue Sharing Agreement between Merced County and the six area cities	None – though letters were sent to the Board of Supervisors and public comment favored the option of growth in already developed areas, decision-makers ultimately selected a development option that was not found by the HIA to be the healthiest	No; Elements 3 (recommendations/mitigations), 4 (monitoring plan), and 5 (transparent documentation) missing; input of stakeholders not solicited in HIA (Element 2.2); magnitude, distribution, and permanence of impacts not assessed (Element 2.4); assumptions, strengths, limits not identified (Element 2.5)
24	*SE 122nd Avenue Planning Study Health Impact Assessment*; 2011; Portland, Oregon	Evaluate both the health impacts of the SE 122nd Avenue Pilot Project recommendations themselves, as well as the health impacts of the 20-minute neighborhood form	Literature review, community consultation, special collection (demographics analysis, field visits/site observations, surveys, modeling, GIS)	Health, environmental	Physical activity, land use, mobility/ access to services, parks and open spaces, air quality, safety and security, social capital, nutrition	Likelihood of impacts (only intended/likely impacts addressed); Direct and indirect impacts; Direction of impacts (positive health impacts, negative health impacts); Magnitude of impacts (portion of population affected); Distribution/equity of impacts (impact on vulnerable populations; groups most likely to be adversely impacted by exposure to outdoor air toxics)	Found that most of the pilot study's recommendations would directly or indirectly improve physical activity, bicycle/ pedestrian safety, and social cohesion (and vulnerable groups would generally share in these positive benefits), but also have the potential for both positive and negative impacts to exposure to outdoor air pollutants and food access. Recommendations included prioritizing improvements in pedestrian and bicycle infrastructure/connectivity; involving immigrant groups and communities of color in designing/improving public spaces; addressing concerns of low-income/minority transit riders; developing a program to monitor changes in outdoor air toxics; conduct a Community Food Assessment; identifying and recruiting business that provide gathering space and healthy food retail; developing a "healthy food zone" ordinance, etc	Undetermined, but general effectiveness assumed at a minimum – efforts were made to integrate health information and stakeholders into the pilot project prior to the HIA	No; Element 4 (monitoring plan) is missing; and permanence of impacts (Element 2.4) not assessed

Table G-3. Continued

ID	HIA; Year; Location	Scope/Summary	Source of Evidence	Impacts/ Endpoints	Pathway of Impact	Characterization of Impact	Decision-making Outcome	Effectiveness of HIA	Minimum Elements of HIA Met? If no, what's missing
26	*Health Impact Assessment: Hawai'i County Agriculture Development Plan*; 2012; Hawaii County, Hawaii	Evaluate the potential positive and negative impacts of three Agriculture Plan policies - institutional buying (farm-to-school programs), commercial expansion of food agriculture, and home production -on the health of Hawaii Island residents	Literature review, community consultation, policy review, special collection (surveys, expert consultation, accessing unpublished data, modeling, focus groups)	Health, economic, environmental	Nutrition, food-borne illness, community/household economics, cultural connectedness, environmental stewardship/ecosystem sustainability	Direction of impacts (positive health impacts, negative health impacts); Magnitude and Likelihood of impacts (based on policy implementation - negligible, small impact on few, moderate impact on many or strong impact on few, strong impact on few or small impact on many, strong impact on many); Distribution/equity of impacts (populations most affected and disproportionate burden of obesity); Quality of evidence	Found that expansion of Farm-to-School programs would improve food security and nutritional quality and create jobs in agriculture/food processing; increased production of fresh food for the local market would improve community food security and nutritional quality and create jobs, increase farm output, and increase farm earnings/state tax revenues; and promotion of home gardening would have a large impact on improving food security and nutrition security (particularly among low-income Hawaii County residents), increase consumption of fruit and vegetables, increase physical activity, and improve individual well-being and community cultural connectedness Provided a number of recommendations for each Agricultural Plan policy, but highlighted two recommendations: expanding Hawaii Island food production so that 30% of food demand can be supplied locally by 2020; and promoting and supporting educational programs for agricultural industry participants	Direct effectiveness – staff members met with the Office of Planning during the state g development period and made additional specific suggestions during the draft review period; all of the key issues and most of the HIA recommendations were included in the final state strategy documents	Yes
29	*Case Study: Bloomington Xcel Energy Corridor Trail Health Impact Assessment*; 2008; Bloomington, Minnesota	Assess potential health impacts and obstacles to the proposed Xcel recreational trail corridor and support for including the Xcel trail corridor in the Alternative Transportation Plan	Community consultation	Health, environmental	Safety and security, mobility/access to services, social capital, physical activity, water quality, air quality, land use	Direction of impacts (positive health impacts); Distribution/ equity of impacts (sectors of the community that would utilize or be affected most by the trail)	Found that the Xcel powerline corridor was suitable for use as a trail corridor and that impacts on health would be positive Recommended enhancements: safety measures (e g, lighting, and police presence); amenities (e g, benches, bathrooms, bike facilities, and quiet spaces); landscape design (e g, community gardens and vegetation buffers); community involvement initiatives; traffic enhancements (e g, traffic-calming measures; and other actions, such as funding and drawing on existing community trail examples	General effectiveness – raised awareness among policy makers; fully documented the results of the HIA workshop outcomes in the Alternative Transportation Plan	No; Elements 2 3 (baseline health conditions), 2 5 (documentation of data sources, assumptions, strengths, limitations) and 4 (monitoring plan) missing; and magnitude, likelihood, and permanence of impacts not assessed (Element 2 4)

Table G-3. Continued

ID	HIA; Year; Location	Scope/Summary	Source of Evidence	Impacts/ Endpoints	Pathway of Impact	Characterization of Impact	Decision-making Outcome	Effectiveness of HIA	Minimum Elements of HIA Met? If no, what's missing
31	*Health Impact Assessment for Proposed Coal Mine at Wishbone Hill, Matanuska-Susitna Borough Alaska (DRAFT)*; 2012; Matanuska-Susitna Borough, Alaska	Review potential positive and negative human health impacts related to the proposed Wishbone Hill Mine (WHM) - a surface coal mine located in the Matanuska-Susitna valley near Sutton, Alaska	Literature review, community consultation, policy review, special collection (focus groups, risk assessments, demographics analysis, modeling, GIS)	Health, environmental/ ecosystem, economic, other (social, cultural, spiritual)	Community/house-hold economics, social capital, safety and security, exposure to hazards, noise pollution, air quality, water quality/quantity, soil quality, nutrition, lifestyle, healthcare access, mobility/access to services, cultural/ spiritual, habitat, visual effects, solid waste	Direction of impacts (positive health impacts, negative health impacts); Distribution of impacts (Zone 1, 2, 3, & 4 based on proximity); Equity of impacts (identification of potentially impacted communities that meet CEQ definitions of minority or low-income population; populations susceptible to air pollution); Likelihood of impacts (high, medium, low); Exposure assessment (particulate matter, chemicals of potential concern in air and water, vapors); Toxicity assessment	Found the mine would have both positive and negative health impacts, but significant data gaps existed prohibiting the full fate-transport and social impacts from being quantified These health effect categories required the highest priority attention – exposure to hazardous materials; water and sanitation; social determinants of health; food, nutrition, and subsistence activity; non-communicable disease; infectious disease; and accidents/injuries Mitigation recommendations, included: a review of best practices; a stakeholder engagement/ communications strategy; air permit requirements for fugitive dust control and monitoring; expanded exposure receptor grid modeling and modeling/monitoring of deposition mode; diesel exhaust particulates, groundwater/surface water monitoring; a transport safety study/risk analysis for major routes; medical emergency response plans and drills; traditional/local knowledge surveys; review subsistence activities in the area; monitor water quality and quantity effects; and monitor local hospital emergency room visits and discharge data	Undetermined – public comments collected on draft HIA and are being incorporated into the final HIA; it appears no further steps have been taken to move forward on the coal mine (e g , the required air permit application), but the Matanuska Susitna Borough Assembly ignored the requests of local health professionals, coalition partners and concerned citizens asking them to pass a resolution that would require a Comprehensive Health Impact Assessment	No; Element 4 (monitoring plan) missing; magnitude and permanence of impacts not addressed (Element 2 4); and funding sources not transparent (Element 5)
34	*City of Ramsey Health Impact Assessment*; 2008; Ramsey, Minnesota	Assess the potential health impacts of current city planning practices, set goals for improvement, and develop future policy directions in conjunction with the 2008 City of Ramsey Comprehensive Plan update	Literature review, policy review, special collection (threshold scoring, GIS)	Health, environmental, services	Air quality, exposure to hazards, land use, housing, mental health, mobility/ access to services, parks and recreation, nutrition, safety and security, social capital, water quality, physical activity	Impacts not characterized/ judged - only current achievement of thresholds scored	Found that the best scoring indicators were for air quality, housing quality, mental health, safety, and social capital; access to transit/densities to support transit, close retail/supermarket opportunities, urban services, and an extensive sidewalk/trail system were limited Recommended future planning efforts focus on areas that are realistic to improve upon by policy makers for city development (i e , land use planning to improve the city's health threshold measures) Recommendations included items such as securing a stop on the Northstar commuter rail, implementing zoning changes, requiring tree canopy and buffer zones, adopting a Complete Streets policy, etc	Direct effectiveness – recommendations from the HIA were used to provide many Goals and Strategies within various chapters of the Comprehensive Plan	No; Elements 2 1 (scoping), 2 2 (stakeholder input), 2 3 (baseline health conditions) 2 4 (judgement of impacts), and 4 (monitoring plan) are missing; and HIA is not transparent (Element 5)

Table G-3. Continued

ID	HIA; Year; Location	Scope/Summary	Source of Evidence	Impacts/ Endpoints	Pathway of Impact	Characterization of Impact	Decision-making Outcome	Effectiveness of HIA	Minimum Elements of HIA Met? If no, what's missing
37	*Atlanta Beltline Health Impact Assessment*; 2007; Atlanta, Georgia	Make health a part of the Atlanta BeltLine decision-making process by predicting health consequences, informing decision makers and the public about health impacts, and providing realistic recommendations to prevent or mitigate negative health outcomes	Literature review, policy review, special collection (surveys, demographics analysis, modeling [forecasting], GIS)	Health, environmental, behavioral, economic, infrastructure, services, demographic, other	Air quality, community/house-hold economics, exposure to hazards, land use, housing, lifestyle, noise pollution, parks and recreation, mental health, mobility/ access to services, nutrition, public health services, safety and security, social capital, soil quality, physical activity, water quality	Direction of impacts (positive health impacts, negative health impacts, neutral health impacts); Magnitude of impacts (affected populations identified for each impact; impact on number of individuals living, working, and going to school in proximity to the TAD; larger city and regional impacts); Distribution/ equity of impacts (demographic and geographic analysis to determine equity of impacts; census tracts with the highest concentration of vulnerable populations identified); Permanence/severity of impacts (identified impacts with the most serious potential health consequences); Timeline of impacts	Identified several issues that transcend specific health impacts and are more due to the challenges of implementing a large, multi-faceted project: timing of various components of the BeltLine, integration of the BeltLine, prioritization of people, design that accommodates all users, and processes that substantively involve all stakeholders and coordinate efforts Identified recommendations to address the overarching issues, as well as impacts on access and social equity, physical activity, safety, social capital, and the environment Prioritized recommendations included items such as making health protection/promotion a consideration in public funding priorities and timing; Safe Routes to Schools programs; a coordinated fare/schedule system for transit; a 25-year public involvement plan; a single information hub; adding park acres to meet the target of 10 acres/1,000 people; improvements to trail accessibility; policies/programs to prevent displacement; access to healthy foods in the southeast planning area; educational intervention to encourage physical activity; and locating sensitive uses away from high-volume road segments or mitigating air pollution	Undetermined	Yes

Table G-3. Continued

ID	HIA; Year; Location	Scope/Summary	Source of Evidence	Impacts/ Endpoints	Pathway of Impact	Characterization of Impact	Decision-making Outcome	Effectiveness of HIA	Minimum Elements of HIA Met? If no, what's missing
38	*Zoning for a Healthy Baltimore: A Health Impact Assessment of the TransForm Baltimore Zoning Code Rewrite*; 2010; Baltimore, Maryland	Evaluate the impacts of Baltimore's comprehensive zoning code re-write, TransForm Baltimore, to maximize the potential for the zoning recode to prevent obesity and other adverse health outcomes and reduce inequities in these outcomes among children and adolescents in Baltimore	Literature review, community consultation, policy review, special collection (interviews, expert consultation, GIS)	Health, behavioral, infrastructure	Land use, lifestyle, mobility/access to services, nutrition, physical activity, safety and security	Direction of impacts (positive health impacts, negative health impacts, unclear health impacts); Magnitude of impacts (percent increase in residents living in neighborhoods and districts that meet certain parameters under the new draft code); Distribution/equity of impacts (identified likelihood of impacts for residents based on poverty level)	Identified elements of the draft zoning code that should remain in the final version of the code (i.e., supported elements), elements that should be revised to promote health and welfare and mitigate the unintended negative health consequences, and recommendations for the code rewrite process and planned administration of the new code (once ratified). Supported elements were those that improved access to healthy food, created walkable environments, strengthened the link between health and zoning, and were easy to use. Recommended revisions to the draft code included: preventing concentration of off-premise alcohol sales outlets and address problematic existing off-premise alcohol sales outlets; creating a walkable environment by including CPTED; and developing incentives for Healthy Food Stores through the zoning code and through other mechanisms. Changes to the code re-write process and plan for code administration included recommendations for incorporating stakeholder feedback and enhancing public engagement, conducting mapping meetings, and other strategies to make the new zoning code as easy to use as possible	Undetermined	No; Element 4 (monitoring plan) missing; and likelihood and permanence of impacts not assessed (Element 2.4)
39	*Hood River County Health Department Health Impact Assessment for the Barrett Property*; 2011; Hood River County, Oregon	Investigate the potential health benefits of turning a former orchard into a community park with open play fields, trails, and community gardens and the potential health risks for users of the property from exposure to residual pesticide chemicals	Literature review, community consultation, policy review, special collection (interviews, surveys, focus group, demographics analysis, GIS)	Health, behavioral, economic, environmental	Exposure to hazards, land use, lifestyle, nutrition, parks and recreation, physical activity, social capital, soil quality, water quality, noise pollution, safety and security, mobility/access to services, community/household economics	Direction of impacts (positive health impacts, negative health impacts); Magnitude of impacts (identified those affected by the development); Distribution/ equity of impacts (impact on vulnerable populations)	Found the Hood River County health-related needs (including chronic disease management and risk factors, nutrition and food insecurity, and behavioral and social health) could be addressed with a park on the Barrett Property. The HIA made recommendations to maximize the health impact of the Barrett Property, including: grading and preparing the land for park development; testing the soil to determine potential chemical residues present on the land from previous pesticide use; monitoring for unintended health and cost consequences; designing the layout of the park to take into consideration the desired use of the land by the entire community, particularly vulnerable populations; once developed, promote availability and accessibility of the park to vulnerable populations; attract wellness programming/events to the park; and educate decision-makers about the HIA	Undetermined – park is still being planned; in June 2012, there was an appeal by orchardists to block the building of the park	No; likelihood and permanence of impacts not assessed (Element 2.4); and limitations and uncertainties not identified (Element 2.5)

Table G-3. Continued

ID	HIA; Year; Location	Scope/Summary	Source of Evidence	Impacts/ Endpoints	Pathway of Impact	Characterization of Impact	Decision-making Outcome	Effectiveness of HIA	Minimum Elements of HIA Met? If no, what's missing
41	Technical Report 9: Sub-Highway 99 Sub-Area Plan Health Impact Assessment; Clark County, Washington; Unknown	Support the Sub-Area Plan vision (to apply land use planning to build a healthy community) by using an established socio-ecological model of health promotion to validate the plan's health promoting features	Literature review, policy review, special collection (GIS)	Health, environmental/ ecosystem, behavioral, economic, infrastructure, services	Air quality, community/house-hold economics, mental health, education, housing, land use, lifestyle, mobility/ access to services, noise pollution, nutrition, parks and recreation, physical activity, safety and security, social capital	Direction of impacts (positive health impacts, negative health impacts, status quo); Distribution/equity of impacts (general description of "population affected;" impact on vulnerable populations)	Identified the health outcomes of the development if the community vision elements were or were not achieved. A list of recommendations were provided for achieving affordable housing, living wage jobs, mixed income residential areas, mixed use areas, increased air quality, decreased noise pollution, access to healthy foods, urban trees, access to transit, bicyclist and pedestrian friendly areas, parks and green spaces, reduced traffic risks, and community safety	Direct effectiveness – HIA was included in the Appendix of the final Area 99 Sub-Area Plan and policy makers embraced some of the recommendations, such as promoting access to stores and services by locating these developments within walking distances of neighborhoods	No; Element 2.2 (stakeholder input) and 4 (monitoring plan) missing; magnitude, likelihood, and permanence of impacts not assessed (Element 2.4); limitations or uncertainties behind data not identified (Element 2.5); and documentation of funding sources and HIA point-of-contact not transparent, no participants other than "Clark County Health" named, and no context for why the county conducted the HIA (Element 5)
43	Inupiat Health and Proposed Alaskan Oil Development: Results of the First Integrated Health Impact Assessment/ Environmental Impact Statement for Proposed Oil Development on Alaska's North Slope; 2007; National Petroleum Reserve, Alaska	Developed as part of a supplemental Environmental Impact Statement (EIS) to examine health impacts of oil and gas development in the Teshekpuk Lake Special Area of the Northeast National Petroleum Reserve (NPR)-A (North Slope Bureau, Alaska)	Literature review, community consultation, special collection (interviews)	Health, environmental/ ecosystem, behavioral, economic	Air quality, community/house-hold economics, land use, exposure to hazards, infectious disease, lifestyle, mental health, nutrition, safety and security, soil quality, water quality, social pathology	Direction and Likelihood of impacts (definite negative health impacts, probable negative health impacts, speculative negative health impacts, unlikely negative health impacts, positive health impacts); Distribution/equity of impacts (impacts on low-income and Native Americans assessed; risks evaluated in the context of disparate incidence, prevalence, and mortality from cancer); noted extent of impact based on degree of disturbance	Highlighted a number of potential health risks and benefits of the proposed leasing, including: impacts on the local diet and rates of obesity and diabetes; exposure to pollution; social problems; and the use of oil and gas revenues to support local services important to health Mitigation measures recommended by the HIA included establishment of a health advisory board to monitor impacts, public health monitoring, studies and management of fish and game, contaminant control, public safety measures, infectious disease controls, an oil spill control plan, and a sustainable community plan	Direct effectiveness – HIA was included as part of the Supplemental EIS; BLM included HIA mitigation measures that fell within its statutory authority (land management) BLM agreed to consider a measure that would require BLM and developers to work with a Health Advisory Board to further delineate impacts and identify and institute appropriate mitigations	Yes

Table G-3. Continued

ID	HIA; Year; Location	Scope/Summary	Source of Evidence	Impacts/ Endpoints	Pathway of Impact	Characterization of Impact	Decision-making Outcome	Effectiveness of HIA	Minimum Elements of HIA Met? If no, what's missing
44	*Page Avenue Health Impact Assessment*; Unknown (possibly 2010); Pagedale, Missouri	Provide an impartial assessment of the health impacts of the Page Avenue Redevelopment on individuals, youth, and families living primarily in Pagedale, Missouri as well as surrounding communities in University City and Wellston	Literature review, community consultation, special collection (interviews, focus surveys, focus groups, risk assessment, GIS)	Health, behavioral, economic, infrastructure, services	Air quality (indoor), community/house-hold economics, education, exposure to hazards, health-care access, mental health, housing, lifestyle, mobility/ access to services, nutrition, physical activity, public health services, safety and security, social capital, noise pollution (indoor)	Assessed impacts of Redevelopment Plan and each of the top 5 recommendations for the seven priority impacts; Direction of impacts (positive health impacts, negative health impacts, no known impacts, significant health impact); Likelihood of impacts (speculative impacts, probable impacts, definite impacts); Distribution/equity of impacts (populations at greater risk to some adverse health endpoints); Overall population impact/Magnitude of impact (high or moderate; based on expected reach and likelihood of impact across all priority impacts)	Overall, the redevelopment will positively impact the health of the community; the only potential negative impacts concerned relocation of people's homes and businesses Because of concerns about prolonged phasing or lack of detail in the current plan, some priority impacts were more certain (access to goods, services, and recreation; access to healthy foods; housing) than others (employment, pedestrian safety, community safety, community identity) The HIA identified Top Recommendations representing common themes from the assessment (replacing symbols of disinvestment and improving pedestrian infrastructure; implementing orchards and gardens; supplementing physical improvements with education and programming; prioritizing opportunities for youth recreation; and foster stakeholder engagement) and fifty-one (51) specific recommendations in seven priority areas (Employment, Access to Goods, Services & Recreation, Access to Healthy Foods, Pedestrian Safety, Community Safety, Community Identity, Housing)	Undetermined (Note: Publication date not given in document but online research shows that it was published in January 2012 There has probably not been enough time for this to have had an impact)	No; Element 4 (monitoring plan) missing; and permanence of impacts not assessed (Element 2 4)

G-32

Table G-3. Continued

ID	HIA; Year; Location	Scope/Summary	Source of Evidence	Impacts/ Endpoints	Pathway of Impact	Characterization of Impact	Decision-making Outcome	Effectiveness of HIA	Minimum Elements of HIA Met? If no, what's missing
45	*Pittsburg Railroad Avenue Specific Plan Health Impact Assessment*; 2008; Pittsburg, California	Determine the health impacts of the Pittsburg Railroad Avenue Specific Plan - a transit-oriented design plan to build a new train station, new residential and commercial uses, public space, and pedestrian and bicycle improvements	Literature review, community consultation, special collection (interviews, demographics analysis, HDMT, modeling, GIS)	Health, behavioral, economic, infrastructure, services, demographic, environmental	Air quality, community/household economics, land use, healthcare access/insurance, education, exposure to hazards, housing, lifestyle, mental health, mobility, access to services, noise pollution, nutrition, parks and recreation, physical activity, safety and security	Positive and negative impacts shown for all pathways and endpoints except for noise, which had only negative impacts; Direction and Likelihood of impacts (definite positive/negative health impacts, probable positive/negative health impacts, speculative positive/negative health impacts); Distribution/equity of impacts (populations more vulnerable to vehicle collisions and air pollution); Quantification of impacts (neighborhood completeness [HDMT]; vehicle trip generation and greenhouse gas emissions [URBEMIS and EMFAC]; PM2.5 residential use- and traffic-related pollution emissions [CALINE3QHCR]; % change in health endpoints due to PM2.5 concentrations); Permanence of impacts (PM2.5 pollution effects high, medium, low; accidents and injuries (by speed); Magnitude of impacts (PM2.5 population impacts)	Identified opportunities that would improve health through the creation of a complete neighborhood around the BART station, as well as some modifiable health and environmental quality threats associated with the project's location adjacent to a freeway corridor. Forty-five (45) recommendations were made by the HIA to increase benefits of the plan and to mitigate negative impacts, including conducting a retail and public services needs assessment; allotting more affordable housing in the project than the current zoning ordinance called for; including high-quality ventilation systems in any housing within one-half mile of the freeway; installing triple-paned windows to protect from noise; hiring local residents for the construction phase of the project; implementing traffic calming measures; implementing strategies to encourage use of BART and decreased use of cars; and locating residential uses and other sensitive land uses in the project area to minimize exposure to significant sources of air pollution and noise	Direct effectiveness – the Planning Department used results from the HIA to save affordable housing sites originally facing opposition, require air quality and noise mitigation measures, and improve pedestrian and bicycling facilities; HIA process also engaged community residents in data collection and partnered with a local health clinic	No; Element 4 (monitoring plan) missing
47	*The East Bay Greenway Health Impact Assessment*; 2007; Oakland to Hayward, California	Highlight potential positive impacts of the Greenway pedestrian and bike trail could have on health and to uncover and suggest mitigations for potential barriers that would hinder the project from reaching its full positive health impact	Literature review, community consultation, special collection (GIS)	Health, behavioral, infrastructure	Land use, lifestyle, mental health, mobility/access to services, parks and recreation, physical activity, safety and security, social capital, air quality, noise pollution	Direction of impacts (positive health impacts, negative health impacts); Distribution/equity of impacts (demographics of affected communities)	The Greenway, as proposed, presents an opportunity in land use that could be very beneficial to the health of residents who live near the route, many of whom are poor, are people of color, and currently suffer from health inequities. Recommended building the Greenway with specific design features (e.g., connecting to existing trails/paths, universal design principles), offering programming to maximize usage, and implementing mitigation steps to increase safety (e.g., lighting, police involvement, traffic calming, etc.)	Direct effectiveness – HIA was included as an appendix to the final East Bay Greenway Concept Plan and many of the recommendations were built into the plan itself	No; Element 4 (monitoring plan) missing; and magnitude, likelihood, and permanence of impacts not assessed (Element 2.4)

Table G-3. Continued

ID	HIA; Year; Location	Scope/Summary	Source of Evidence	Impacts/ Endpoints	Pathway of Impact	Characterization of Impact	Decision-making Outcome	Effectiveness of HIA	Minimum Elements of HIA Met? If no, what's missing
49	*Taylor Energy Center Health Impact Assessment;* 2007; Taylor County, Florida	Analyze the impact of a proposed coal-fired electric plant, including risks from air pollution and benefits to health from employment by the plant and the "community contribution"	Literature review, community consultation, policy review, special collection (risk assessment, demographics analysis, modeling)	Health, economic, environmental/ ecosystem	Air quality, community/ house-hold economics, exposure to hazards, infectious disease, land use, water quality	Direction and Likelihood of impacts (definite negative health impacts, definite positive health impacts, speculative positive health impacts, definite no impact); Magnitude of impacts (low, medium, high); Distribution/equity of impacts (racial disparities in health and sensitive populations identified; change in risk of mortality due income); Permanence and Quantification of impacts (impacts on life expectancy [additional or averted death estimates])	Found substantial racial disparities in health and predicted both positive (economic) and negative (air pollution) health impacts from the Taylor Energy Center. Provided multiple recommendations, including establishing baseline levels/ monitoring mercury emissions; installing an air quality monitor; adopting a policy to remain carbon negative; targeting job recruitment to include a representative or greater proportion of black residents; recruiting and training a diverse population of Taylor County residents for professional jobs at TEC; and investing in the community. Also provided one recommendation to mitigate the smoking attributable mortality discovered in the baseline health status – to implement additional smoking cessation programs and provide health prevention/education programs	Undetermined – no mention of HIA on the official project site; however, they do appear to be doing some air quality monitoring as recommended by the HIA	Yes
58	*Battlement Mesa Health Impact Assessment (2nd Draft);* 2010; Battlement Mesa, Colorado	Address citizen concerns about health impacts of natural gas development and production in the Battlement Mesa Planned Unit Development (PUD)	Literature review, policy review, special collection (demographics analysis, risk assessment, GIS)	Health, environmental, behavioral, economic, infrastructure, services, demographic	Air quality, community/ house-hold economics, education, exposure to hazards, land use, healthcare access/insurance, housing, infectious disease, lifestyle, mental health, noise pollution, safety and security, social capital, soil quality, water quality	Direction of impacts (negative health impacts, positive health impacts); Permanence of impacts (duration of exposure long, short to long); Magnitude of impacts (moderate to high, low to high, low to medium, low); Likelihood of impacts (likely, possible, unlikely); Distribution/geographic extent of impacts (local, community-wide); Equity of impacts (vulnerable populations identified); Frequency of impacts (infrequent, frequent, or constant)	Found that the health of the Battlement Mesa residents would most likely be affected by chemical exposures, accidents, or emergencies resulting from industry operations, and stress-related community changes. General recommendations in the HIA included pollution prevention, protection of public safety, and increased communication through the development of a Community Advisory Board. Over 70 specific recommendations were also provided in the eight areas of concern identified by the community. These recommendations focused on reducing air emissions, continued monitoring of air and water sheds, and strict enforcement of existing regulations; use of best available current technology and rapid adoption of new technologies to decrease emissions; reduction of risk of traffic and industrial accidents; and development of a community advisory board to facilitate communication with the goal of improving community well-being	Direct effectiveness – the Board of County Commissioners recognized the many gaps and monitoring needs that were identified in the HIA and contracted with the Colorado School of Public Health to design an Environmental and Health Monitoring Study (EHMS) that could begin to gather this information; the HIA will be used as a reference document if further land use application is made	Yes

Table G-3. Continued

ID	HIA; Year; Location	Scope/Summary	Source of Evidence	Impacts/ Endpoints	Pathway of Impact	Characterization of Impact	Decision-making Outcome	Effectiveness of HIA	Minimum Elements of HIA Met? If no, what's missing
59	*Douglas County Comprehensive Plan Update Health Impact Assessment*; 2011; Douglas County, Minnesota	Evaluate updates to the Douglas County Comprehensive Plan, which provides a framework and policy direction for future land use, transportation, natural resource and park/open space decisions	Policy review, community consultation, special collection (demographics analysis, GIS)	Health, behavioral, economic, infrastructure, services	Healthcare access, mental health, mobility/access to services, parks and recreation, physical activity, safety and security, social capital, land use	Health indicators used to measure health issues - provided a description of the indicator, supporting language and policy statements already in the Plan that address the indicator, and recommendations for language and/or policy statements to be added to the Plan; Distribution/ equity of impacts (assessed impacts on the aging population)	The Comprehensive Plan was assessed against 12 health indicators - identifying aging population and senior services, connectivity, recreational amenities (community facilities, gardens, parks, and trails), economic opportunities, mixed-use development, traffic accidents, and complete streets/traffic calming. Specific language and policy statements were recommended for incorporation into the Final Comprehensive Plan in order to ensure that priority health areas of concern were addressed	Direct effectiveness – several language and policy recommendations were included in the final plan	No; Elements 2, 3 (baseline health conditions, except for baseline aging conditions) and 4 (monitoring plan) missing; direction, likelihood, magnitude, and permanence of impacts not assessed (Element 2, 4); and little supporting evidence used/documented (Element 2, 5)
60	*The Executive Park Subarea Plan Health Impact Assessment: An Application of the Healthy Development Measurement Tool (HDMT)*; 2007; San Francisco, California	Summarize the results from the first application of San Francisco's Healthy Development Measurement Tool to the Executive Park Subarea Plan, which proposes to build 2,800 units of new residential housing on a 71-acre area in the southeastern corner of San Francisco	Literature review, policy review, special collection (demographics analysis, modeling, GIS, HDMT)	Health, environmental/ ecosystem, behavioral, economic, infrastructure, services	Air quality, education, exposure to hazards, land use, healthcare access/insurance, housing, lifestyle, mental health, mobility/access to services, noise pollution, nutrition, physical activity, public health services, parks and recreation, safety and security, social capital	Direction of impacts (positive health impacts, negative health impacts); Likelihood of impacts (increased likelihood, decreased likelihood); Distribution of impacts (income inequality; potential reduction in existing disparities in health status)	The Executive Park Subarea Plan met between one-third and two-thirds of the development targets for each of the six elements evaluated and overall, approximately 50% of the targets evaluated. In addition to specific recommendations for all of the HDMT objectives, the HIA also provided a number of general/crosscutting recommendations, including additional implementation actions/strategies for incorporation into the plan and improvements to transportation and access to goods and services. Overall, the application of the Healthy Development Measurement Tool (HDMT) to the Executive Park Subarea Plan demonstrated that the HMDT is a feasible methodology that can be used to conduct a comprehensive health and sustainability assessment of land use development projects	Undetermined	No; Element 5 (transparent documentation) missing; permanence and magnitude of impacts not assessed (Element 2, 4); supporting evidence for the HDMT indicators/methodology and references not provided in most cases (Element 2, 5); follow-up measures included in recommendations, but monitoring plan not identified (Element 4)

G-35

Table G-3. Continued

ID	HIA; Year; Location	Scope/Summary	Source of Evidence	Impacts/ Endpoints	Pathway of Impact	Characterization of Impact	Decision-making Outcome	Effectiveness of HIA	Minimum Elements of HIA Met? If no, what's missing
63	*Oak to Ninth Avenue Health Impact Assessment;* 2006; Oakland, California	Assess the influence of the Oak to Ninth Avenue development project – a waterfront mixed-use neighborhood - on determinants of human health	Literature review, policy review, special collection (survey, modeling, GIS)	Health, environmental/ ecosystem, behavioral, economic, infrastructure, services, demographic	Air quality, community/ house-hold economics, education, exposure to hazards, housing, land use, lifestyle, mental health, mobility/access to services, noise pollution, parks and recreation, physical activity, safety and security, social capital	Direction of impacts (positive health impacts, negative health effects); Magnitude of health impacts (additional traffic related injuries per year; % residents that will experience sleep disturbance; health effects of freeway air pollutants over 10 years); Distribution/equity of impacts (populations affected by lack of affordable housing, disparities in park accessibility, air quality, noise, and pedestrian safety); Quantification of impacts (forecasted changes to pedestrian injury rates; traffic-related emissions)	Estimated a number of negative health impacts from the development and provided recommendations related to each factor evaluated. Major recommendations/mitigations included: improving public/stakeholder participation, improved access to the waterfront, traffic calming and other pedestrian safety measures, distribution of housing costs and mixed-income housing, public transit and other options for reducing VMT, in-home systems for air quality and noise mitigation (HVAC, insulating windows, etc)	Direct effectiveness – the project's environmental impact report (EIR) was found deficient by a California Superior Court, resulting in invalidation of the Oakland City Council's adoption of that EIR and related documents; however, the EIR was revised and in January 2009, the Oakland City Council adopted a resolution to approve the EIR revisions and re-adopt the related EIR certification	No; Element 4 (monitoring plan) missing; baseline conditions not well-established (Element 2 3); and likelihood and permanence of impacts not assessed (Element 2 4)
64	*A Health Assessment of Mixed Use Redevelopment Nodes and Corridors in Lincoln, Nebraska;* 2011; Lincoln, Nebraska	Analyze the nodes and corridors proposal in the Comprehensive Plan to determine whether the proposed changes would truly generate health benefits in Lincoln	Literature review, policy review, special collection (modeling, stakeholder meeting, GIS)	Health, environmental/ ecosystem, behavioral	Land use, physical activity, air quality, mobility/access to services	Direction of impacts (positive health impacts); Distribution/ equity of impacts (benefits attributable to the Lincoln population; populations most affected by improved air conditions); Quantification of impacts (change in walkability)	The walkability analysis showed mixed use redevelopment in the identified nodes will generate an increase in walkability over current conditions and a subsequent improvement in air quality due to reduced vehicle travel. No recommendations provided	None	No; Elements 3 (recommendations and mitigations), and 4 (monitoring plan) missing; identification of baseline conditions limited (Element 2 3); and not all potential impacts assessed (Element 2 4)

G-36

Table G-3. Continued

ID	HIA; Year; Location	Scope/Summary	Source of Evidence	Impacts/ Endpoints	Pathway of Impact	Characterization of Impact	Decision-making Outcome	Effectiveness of HIA	Minimum Elements of HIA Met? If no, what's missing
67	*Healthy Tumalo Community Plan: A Health Impact Assessment on the Tumalo Community Plan; A Chapter Of The 20-Year Deschutes County Comprehensive Plan Update*; 2010; Tumalo, Oregon	Evaluate the draft Tumalo Community Plan in the context of community health and support county planners by providing recommendations that could be incorporated into the final plan	Literature review, community consultation, special collection (community meetings, surveys, GIS)	Health, infrastructure, behavioral	Physical activity, safety and security, rural livability, social capital, mobility/access to services, parks and open space	Direction of impacts (negative health impacts, positive health impacts); Distribution/equity of impacts (vulnerable populations at risk of obesity)	Recommended changes to existing policies in the Tumalo Community Plan and/or the addition of new policies to promote positive health outcomes, including improving the safety and accessibility of the major highway that runs through town, creating new parks and infrastructure to maximize the safe and healthy use of riverfront property as a recreational facility, and building trails or other connections between existing recreational facilities and downtown, local schools and businesses	Direct effectiveness – some community input from HIA process incorporated into updated Tumalo Community Plan language, and a needs assessment is also currently underway as a result of the HIA project to develop a Safe Routes to School program	No; baseline health conditions not well established (Element 2.3); magnitude, likelihood, and permanence of impacts not assessed (Element 2.4); sources of data and synthesis of evidence not transparent (Element 2.5); and no plans for monitoring provided (Element 4)
68	*Strategic Health Impact Assessment on Wind Energy Development in Oregon (Public Review Draft)*; 2012; Oregon	Assess ways that wind energy developments in Oregon might affect the health of individuals and communities where they are built and maintained, develop evidence-based recommendations for future facility siting decisions, engage community and stakeholders, and assess the utility of HIA for specific wind farm siting decisions	Literature review, policy review, community consultation, special collection (survey, GIS)	Health, economic, environmental	Air quality, climate change, community/household economics, education, exposure to hazards, infectious disease, land use, noise pollution, safety and security, social capital	Direction of impacts (positive health impacts, negative health impacts); Likelihood of impacts (definite impacts, possible impacts, unlikely impacts); Permanence and Magnitude of impacts (low); Distribution/ equity of impacts (populations vulnerable to air pollution effects, night-time noise, wind turbine sound, and community-level conflicts; extent of exposure from construction-related emissions by population; disparities in SES between rural and urban areas); Quantification of impacts (impacts of background sound and long-term outdoor community sound at certain levels)	HIA was not focused on a specific facility or community, but rather what is currently known about the health impacts from wind farms (noise, visual impacts, air pollution, economic effects, and community conflict) and the policies and standards used to site wind facilities in Oregon. The HIA was designed to provide a framework and reference materials for future assessments and decisions on proposed wind energy installations. Provided several general recommendations, such as implementing strategies to minimize sound generation; addressing community concerns as part of the siting process; using up-to-date, current state of science noise modeling to plan facilities boundaries and turbine locations; considering the distance, orientation and placement of turbines relative to homes and buildings to reduce shadow flicker; and increasing community-wide economic benefits from wind energy developments. Also provided recommendations and mitigation strategies for site-specific wind facility assessments, including tools and models for assessing baseline air pollutant levels and local air pollution impacts; systems and protocols for documenting, responding to, and evaluating complaints; implementing sound mitigation strategies; and using visual obstructions to block flicker	Undetermined	No; baseline health conditions not established (Element 2.3)

Table G-3. Continued

ID	HIA; Year; Location	Scope/Summary	Source of Evidence	Impacts/ Endpoints	Pathway of Impact	Characterization of Impact	Decision-making Outcome	Effectiveness of HIA	Minimum Elements of HIA Met? If no, what's missing
69	*Impacts on Community Health of Area Plans for the Mission, East SoMa, and Potrero Hill/Showplace Square: An Application of the Healthy Development Measurement Tool;* 2008; San Francisco, California	Use the Healthy Development Measurement Tool (HDMT) to examine potential health implications of the Eastern Neighborhoods Area Plans, using 26 of 27 community health objectives within six healthy city vision elements – environmental stewardship, sustainable and safe transportation, social cohesion, public infrastructure/access to goods and services, adequate and healthy housing, and healthy economy	Literature review, policy review, special collection (HDMT, modeling)	Health, environmental/ ecosystem, behavioral, economic, infrastructure, services	Environmental stewardship, air quality, community /household eco-nomics, education, exposure to hazards, land use, healthcare access/insurance, housing, lifestyle, mental health, mobility/access to services, noise pollution, nutrition, parks and recreation, physical activity, public health services, safety and security, social capital	Quantification (HDMT used to evaluate whether Area Plans met Development Targets (proxies for meeting Community Health Objectives); Direction of impacts/plan strengths and weaknesses (positive health impacts/plan strengths, negative health impacts/plan weaknesses); Distribution/equity of impacts (populations vulnerable to air pollution; disparities in proximity of households to roadway air pollution sources; populations with limited ability to walk)	Eastern Neighborhood Area Plans met approximately 55% of the analyzable development targets (i.e., targets that were applicable and for which adequate data was available) Identified several concerns revolving around Plan implementation and collaboration and deferral of Area Plan implementing actions to future studies Based on HDMT evaluation, a number of recommendations were developed that related to environmental sustainability, preservation of open space, community-supported agriculture (CSA) and community gardens, identification of pollution sources, reviews of zoning to minimize locating sensitive uses in close proximity to those sources, traffic calming measures, etc Overarching recommendations for comprehensive community planning were also developed, such as increased specificity/ level of detail in the Area Plans and transparency of how decisions regarding Plan and rezoning elements are made, and timely coordination of studies informing the Area Plans	Direct effectiveness – SFDPH provided input on planning strategies to meet identified health needs and policy and implementation recommendations at several stages of the Area Plan development (including during evaluation of the draft and final Area Plans using the HDMT); some of these recommendations were incorporated	No; Element 4 (monitoring plan) missing; baseline conditions for health not documented (Element 2, 3); magnitude, likelihood, distribution, and permanence of impacts not assessed (Element 2, 4); sources of data not acknowledged and synthesis of information not transparent (Element 2, 5); and documentation not transparent (Element 5)

Table G-3. Continued

ID	HIA; Year; Location	Scope/Summary	Source of Evidence	Impacts/Endpoints	Pathway of Impact	Characterization of Impact	Decision-making Outcome	Effectiveness of HIA	Minimum Elements of HIA Met? If no, what's missing
71	*MacArthur BART Transit Village Health Impact Assessment*; 2007; Oakland, California	Examine the health impacts of the MacArthur BART Transit Village - a proposed redevelopment of the MacArthur Bay Area Rapid Transit Station parking lot and adjacent property into a mixed use village	Literature review, community consultation, policy review, special collection (survey, field visits, modeling, demographics analysis, GIS)	Health, environmental, economic, services	Housing, mobility/ access to services, parks and recreation, safety and security, air quality, noise pollution, social capital; secondary pathways: house-hold/community economics, physical activity, exposure to hazards, nutrition	Direction of impacts (positive health impacts, negative health impacts); Likelihood of impacts (definite impacts, probable impact); Magnitude of impacts (number of housing units; regional air quality impacts); Distribution/equity of impacts (park access inequities; affordable housing locations to ensure against environmental injustice; populations vulnerable to pedestrian-vehicle injuries; air pollutants, noise levels; populations disproportionately affected by violent crime; populations that could benefit greatly from open space); Quantification of impacts (student generation estimates; forecast of child care demand and changes to pedestrian injury rate; modeled PM2 5 levels and forecasted health effects; cancer risk estimation due to diesel particulate matter; projected carbon monoxide exposure; measured/modeled noise levels)	Found that the MacArthur Bart Transit Village could impact a large number of individuals, many of low socioeconomic status, both positively and negatively Provided over 80 recommendations, including consideration of basic safety and affordability needs; strategies to meet sustainable transportation goals; essential retail services for the mixed-use retail corridor (e g , full service grocery); ample quality parks and natural space; a comprehensive pedestrian safety countermeasure plan; mitigations to reduce air and noise pollution exposure; incorporating CPTED elements into the design; strategies to include more west side residents in design and planning; unbundling parking from unit sales; bicycle parking and connection to the local bike network; pedestrian safety improvements especially for school routes; and using building materials and ventilation systems to reduce allergens and toxic exposures	Undetermined	No; Element 4 missing (monitoring plan); and permanence of impacts not assessed (Element 2 4)

G-39

Table G-3. Continued

ID	HIA; Year; Location	Scope/Summary	Source of Evidence	Impacts/ Endpoints	Pathway of Impact	Characterization of Impact	Decision-making Outcome	Effectiveness of HIA	Minimum Elements of HIA Met? If no, what's missing
72	*Healthy Corridor for All: A Community Health Impact Assessment of Transit-oriented Development Policy in St. Paul Minnesota*; 2011; St Paul, Minnesota	Examine the rezoning ordinance that would lay the foundation for the implementation of transit-oriented development (TOD) along the Central Corridor to understand the impacts of the light rail line and subsequent land use changes on community health, health inequities, and underlying conditions that determine health	Literature review, special collection (demographics analysis, modeling, risk assessment, survey, GIS)	Health, economic, infrastructure, demographic, services	Community/household economics, housing, exposure to hazards, land use, mobility/ access to services, physical activity, safety and security, social capital	Direction of impacts (positive health impacts, negative health impacts); Distribution/equity of impacts (racial disparities in household income, educational attainment, unemployment; geographic health disparities; populations that would experience increased access, housing burden, and displacement; populations more vulnerable to negative impacts; populations that must be considered and heard); Permanence of some impacts (low or medium/near term, high/long-term/ irreversible); Magnitude of impacts (population affected); Likelihood of impacts (definite impacts, probable impacts, speculative impacts); Quantification of impacts (loss of on-street parking; localized job analysis; statistical analyses; geographical analysis)	Analyzed how the anticipated changes in land use would affect existing conditions in the corridor according to two different scenarios; one was market-based using estimates from a market analysis conducted by a real estate firm and the other used the maximum allowable development outlined in the rezoning proposal The most important finding was the vulnerability of communities of color and low-income individuals in the Central Corridor to the potential negative impacts of the rezoning and new light rail line Best practices in equitable development were identified and recommendations developed to mitigate the negative impacts and maximize the positive outcomes of the zoning Considering local context and community needs, five policy recommendations were prioritized around development and preservation of affordable housing, and developed into more detailed policy briefs Priority recommendations were: developing a community equity program to retain affordable housing; codifying the City's commitment to affordable housing; implementing a density bonus program for developers who provide affordable housing in new residential and mixed-use development projects in the Central Corridor; relieving the lack of commercial parking; and implementing first source hiring	General effectiveness at a minimum, but direct effectiveness assumed – rezoning did not specifically include the priority recommendations of the HIA, but mechanisms were put in place to address the affordable housing issues raised (i e , city council requested feasibility analyses, created a forum for consensus building, and developed a resolution to create an affordable housing work-group); policy debate around the rezoning shifted as a result of the HIA and introduced health into the discussion HIA led to increased capacity building in the community, and HIA steering committee became a coalition engaged in the city zoning decision-making processes	Yes

Table G-3. Continued

ID	HIA; Year; Location	Scope/Summary	Source of Evidence	Impacts/ Endpoints	Pathway of Impact	Characterization of Impact	Decision-making Outcome	Effectiveness of HIA	Minimum Elements of HIA Met? If no, what's missing
73	*Health Impact Assessment Point Thomson Project*; 2011; Alaska	Identify human health impacts associated with each of the five proposed design alternatives of the proposed oil and gas development in Alaska's remote Point Thomson area	Literature review, policy review, community consultation, special collection (interviews, field visits, risk assessment, focus groups, demographics analysis)	Health, behavioral, infrastructure, services, environmental/ ecosystem	Air quality, exposure to hazards, lifestyle, healthcare access, mental health, nutrition, infectious disease, land use, public health services, safety and security, social capital	Direction of impacts (negative health impacts, positive health impact, no health impact); Magnitude (intensity) of impacts (low, medium, high, very high); Duration/frequency (Permanence) of impacts (Less than 1 month/happens rarely; short-term - less than a year/low frequency; medium-term - one to six years/intermittent frequency; long-term - more than six year, life of project/ constant frequency); Distribution of impacts (geographical extent); Likelihood of impacts (exceptionally unlikely, very unlikely, unlikely, about as likely as not, likely, very likely, virtually certain); Nature of impacts (direct, indirect, cumulative); Equity of impacts (potentially affected communities divided into three zones based on likelihood of significant health impacts); Ranked significance of impact/risk assessment score (low, medium, high, very high)	The most significant positive and negative impacts of the project were centered in the Zone 1 communities and the coastal hunting areas utilized by both communities and were associated with transportation corridors; exposures to hazardous materials; emergency medical services; continued evolution of subsistence and nutrition behaviors; and psychosocial effects Mitigation strategies and recommendations developed in response to the medium- to high-impact negative effects and organized around health promotion and disease prevention: 1 Follow proposed EPA regulations on stack emissions and implement baseline stack monitoring; 2 Increase community education about safety measures in place for arctic projects and ongoing community engagement; 3 Restrict access and increase security and safety patrols; 4 Conduct baseline nutritional surveys and ongoing monitoring; 5 Develop response plan for augmentation of existing health care infrastructure in local clinics	General effectiveness – HIA included as Appendix R to the Final EIS and HIA-identified human health impacts noted in the project's Record of Decision (ROD) but the Army Corps of Engineers determined that project would have a minimal detrimental effect on human health; the ROD approved Alternative B (the proposed action) with modifications and mitigation measures	No: Element 4 (monitoring plan) missing; and documentation of funding sources not transparent (Element 5)

Table G-3. Continued

ID	HIA; Year; Location	Scope/Summary	Source of Evidence	Impacts/ Endpoints	Pathway of Impact	Characterization of Impact	Decision-making Outcome	Effectiveness of HIA	Minimum Elements of HIA Met? If no, what's missing
77	*Humboldt County General Plan Update Health Impact Assessment*; 2008; Humboldt County, California	Evaluate six key areas of the Humboldt County General Plan Update (GPU) to identify how indicators of healthy development would change as a result of the three alternatives being considered - denser development in urban areas, limited growth to exurban areas, and unrestricted growth across the county	Literature review, special collection (focus groups, surveys, demographics analysis, GIS)	Health, environmental, infrastructure, economic	Housing, community household economics, land use, mobility/access to services, mental health, nutrition, physical activity, social capital, safety and security, air quality, environmental stewardship	Direction and Likelihood of impacts for each alternative; Distribution/equity of impacts (vulnerable populations; policies that reduce this disparity most); Direction and Likelihood of impacts from the recommended Plan Alternative: Housing (positive health impacts, definite no change); Transportation (positive health impacts); Public Infrastructure (positive health impacts); Economy (no clear direction or likelihood identified); Safety/Social Cohesion (positive health impact, mixed health benefits); Environmental Stewardship (positive health impacts, definite no change, negative health impacts)	The recommended plan alternative was denser development in urban areas, as it is likely to have the most positive overall health impacts and require the fewest health related mitigations For the six areas considered, recommendations were provided for mitigating negative health impacts and promoting positive health impacts Example recommendations by area: Housing (develop policies to encourage affordable housing, establish programs to assist the homeless population); Transportation (develop policies to increase public transit use and encourage walking and biking); Public Infrastructure (increase access to parks, senior centers medical facilities and childcare, provide incentives to grocery stores selling produce); Economy (develop policies to attract and retain industries that provide a living wage, health insurance and workforce education); Safety/Social Cohesion (activities to promote community building, increased emergency preparedness); Environmental stewardship (restrict housing placement to the periphery of agriculturally zoned land, decrease energy consumption, promote consumption of locally-grown food) HIA also led to the development of a rural HDMT	Direct effectiveness – HIA was included as an Appendix to the GPU; the Housing Element of the GPU increased the amount of affordable housing due to the HIAs results and the Transportation element included some of the HIA research and findings in that section	No; Element 4 (monitoring plan) missing; and magnitude and permanence of impacts not assessed (Element 2 4); and documentation of HIA point-of-contact not transparent (Element 5)

G-42

Table G-3. Continued

ID	HIA; Year; Location	Scope/Summary	Source of Evidence	Impacts/ Endpoints	Pathway of Impact	Characterization of Impact	Decision-making Outcome	Effectiveness of HIA	Minimum Elements of HIA Met? If no, what's missing
78	*Rapid Health Impact Assessment: Vancouver Comprehensive Growth Management Plan 2011*; 2011; Vancouver, Washington	Examine the 2011 Vancouver Comprehensive Plan and its impact on two key determinants of health - physical activity and access to healthy food	Literature review, special collection (surveys, demographics, modeling, GIS)	Health, services	Land use, physical activity, mobility/ access to services, safety and security	Direction and likelihood of impacts (Planning changes – all definite positive health impacts; Zoning changes – minimal, but positive impacts probable); Distribution/equity of impacts (Planning changes – equal impacts or positive differential impacts on geographically-focused populations; Zoning changes – unclear, equal impacts, positive/negative differential impacts on populations); Quantification of impacts (walkability index; Walk Score index; Connected Node Ratio; total Floor Area Ratio; bikeway network density; GIS mapping); Strength of evidence (Planning direction changes – some to strong; Zoning changes – moderate)	Concluded that the proposed planning direction changes, policy changes, and zoning changes would likely be beneficial to community health, but need to be implemented through development standards to be effective Several recommendations were given to further improve opportunities for physical activity and access to healthy food: 1 Physical Activity – developing land uses and transportation networks that support physical activity; enhancing connectivity; managing parking to encourage active transportation; improving safety and comfort for pedestrians and bicyclists; increasing use of transportation modes; and reducing disparities in access to physical activity and protecting vulnerable populations; 2 Healthy Food Access – recruiting and retaining healthy food retail; promoting opportunities to grow food in home and community gardens; reducing the availability of unhealthy food options relative to healthy food options; promoting food security; and reducing disparities in food access and protecting vulnerable populations	Undetermined	No; magnitude and permanence of impacts not assessed (Element 2 4)

G-43

Table G-4. Summary Table of Select Data from HIAs in the Waste Management/Site Revitalization Sector

ID	HIA; Year; Location	Scope/Summary	Source of Evidence	Impacts/ Endpoints	Pathway of Impact	Characterization of Impact	Decision-making Outcome	Effectiveness of HIA	Minimum Elements of HIA Met? If no, what's missing
4	Health Impact Assessment of NRMT's Request for a Special Use Permit; 2011; Bernalillo County, New Mexico	Address the health impacts of the proposed dirty materials recovery facility	Community consultation, literature review, policy review, special collection (applicant information, GIS)	Health, environmental, behavioral, economic, demographic, infrastructure, other	Neighborhood livability, traffic congestion, air quality, noise pollution, odor, community/house-hold economics, mental health, exposure to hazards	Distribution/equity of impacts (siting facility in vulnerable community); Direction of impacts (negative health impacts); Likelihood of impacts (probable impacts, speculative impacts)	Recommended denying the requested special use permit; for a relatively modest recycling achievement, the communities would experience significant health burdens, which would likely contribute to the already statistically significant high death rates and shorter life spans of residents and the potential for further environmental degradation	Direct effectiveness – influential in land-use hearings; permit denied by Planning Commission	No; Element 4 (monitoring plan) missing
25	Concord Naval Weapons Station Reuse Project Health Impact Assessment; 2009; Concord, California	Analyze how the alternatives being considered for the CNWS Reuse Project would help realize health and well being benefits or potentially lead to negative health outcomes	Literature review, community consultation, policy review, special collection (focus groups, demographics analysis, modeling, GIS)	Health, economic, environmental, infrastructure, services, demographic	Housing, commun-ity/household economics, parks and open space; secondary impacts: mobility/access to goods and services, physical activity, nutrition, air quality, water quality, noise pollution, social capital, safety and security	Direction of impacts (negative health impacts, positive health impacts); Distribution/equity of impacts (opportunities for ethnically/economically segregated neighborhood; ensure parks contain facilities useable by individuals with limited mobility)	Concluded that the Concentration and Conservation was the healthier of the two alternatives being considered from several perspectives; however, both alternatives were predicted to lead to negative health impacts if mitigations were not implemented. Recommendations included maximizing residential density near the commuter rail station; increasing the amount of affordable housing; adopting a living wage ordinance; adopting local hiring policies; maximizing the land available for parks and open space; promoting public transit and ensuring neighborhoods are walkable and bikeable; encouraging healthy goods and services to be provided on-site via zoning and other mechanisms. Also noted that some mitigations, especially those regarding affordable housing, must be in place before the footprint of development is finalized and the Navy puts the land to auction	Direct effectiveness – advocates used the HIA to win a plan that has significant amounts of land reserved for parks and open space and relatively high density housing; the Final EIR approved by the City Council responded to some of the recommendations, but changed little. Concord City Council voted to move forward with the second most dense land use option proposed and has taken steps to ensure that a significant amount of affordable housing is built at the site	No; Element 4 (monitoring plan) missing; and no quantification of the magnitude, likelihood, or permanence of impacts (Element 2 4) due to limited availability of final options
33	Neenah-Menasha Sewerage Commission Biosolids Storage Facility, Greenville, WI; 2011; Greenville, Wisconsin	Review potential health concerns and propose methods to reduce those risks with the building of a biosolids storage facility	Literature review, community consultation, policy review, special collection (interviews, survey, aerial photo, GIS)	Health, environmental	Air quality, exposure to hazards, infectious disease, land use, mental health, safety and security, soil quality, water quality	Direction of impacts (negative health impacts); Permanence of impacts (life of the proposed facility); Magnitude of impacts (unknown); Likelihood of impacts (probable impacts, unlikely impacts, uncertain impacts); Distribution/equity of impacts (unknown)	Concluded that if biosolids were handled in an appropriate manner and according to regulations, they should not result in a human health hazard. Recommended that the Neenah-Menasha Sewerage Commission track and respond to complaints and that biosolids-related health complaints be monitored so that trends or other indicators of adverse health effects can be recognized and investigated in a timely manner. Public Health staff will monitor health effects using a standardized tool developed by a North Carolina research group for investigating health incidents associated with biosolids applied to land	Undetermined – unclear if plans to build biosolids storage facility was dropped due to the HIA or public pressure	No; Element 2 3 (baseline health conditions) missing; scoping phase solely focused on health (no social or economic parameters considered: Element 2 1); and lack of transparency in funding and participants (Element 5)

G-44

Table G-4. Continued

ID	HIA; Year; Location	Scope/Summary	Source of Evidence	Impacts/ Endpoints	Pathway of Impact	Characterization of Impact	Decision-making Outcome	Effectiveness of HIA	Minimum Elements of HIA Met? If no, what's missing
74	*Assessment of Open Burning Enforcement in La Crosse County*; 2011; La Crosse, Wisconsin	Determine the potential health impacts of creating a uniform open air burning policy within La Crosse County	Literature review, policy review, special collection (survey)	Health, economics, environmental	Household economics, exposure to hazards, safety and security, air quality	Direction of impacts (negative health impacts); Magnitude of impacts (number of La Crosse County residents; Coulee Region river system); Distribution/equity of impacts (populations at the greatest disadvantage for disposing of waste); Economic impacts (offsetting impacts depending on the area of the county)	For municipalities with limited resources for education and enforcement, a detailed burn policy is an opportunity to provide education and guidance for people on solid waste disposal services and the policies surrounding what can be burned. HIA helped to confirm/update burn information that was compiled by the Solid Waste Department in 2010, and data collected through the HIA will help direct education efforts toward the municipalities. Recommendations provided included inquiring with municipal stakeholders how solid waste service decisions are made and what barriers exist, educating the community about solid waste disposal services and burning rules, making municipality solid waste services and schedules readily available on-line, and collecting survey information from fire chiefs that did not respond to the original HIA survey	Undetermined – no county-wide ordinance found, but the HIA did begin a relationship between Health Department and fire and municipal staff for cooperative consideration of quality of life improvements for residents	No; Element 4 (monitoring plan) missing: likelihood and permanence of impacts not assessed (Element 2.4), and documentation of funding sources not transparent (Element 5)

Appendix H –Rules of Engagement Memo

The following is an excerpt from the Healthy Corridor for All: A Community Health Impact Assessment of Transit-oriented Development Policy in St. Paul Minnesota:

Guidelines for Engagement in Healthy Corridor for All HIA

Healthy Corridor for All
Guidelines for Engagement: Goals, Values, Collaboration, and Roles and Responsibilities

Project Description
Work has begun on a new transit line, the Central Corridor Light Rail Transit line (CCLRT), connecting downtown Minneapolis with downtown St. Paul. The CCLRT is a $1 billion transit investment with potentially up to $2 billion investment in local development. The Central Corridor Development Strategy (CCDS) has been developed to guide this investment. The City of St. Paul is currently undergoing a rezoning process along the corridor in order to enable the CCDS. The anticipated timeline for the rezoning to be complete in Spring 2010.

ISAIAH, Take Action Minnesota/Hmong Organizing Project, and PolicyLink are conducting a Health Impact Assessment (HIA) on the proposed Central Corridor rezoning to identify health benefits and burdens associated with the rezoning, and to make recommendations to alleviate negative health impacts stemming from the rezoning.

HIA Definition: Health impact assessment may be defined as a combination of procedures, methods and tools that systematically judges the potential, and sometimes unintended, effects of a policy, plan, program or project on the health of a population and the distribution of those effects within the population. HIA identifies appropriate actions to manage those effects. *(International Association of Impact Assessment, 2006)*

Purpose: To analyze the potential positive and negative health implications of land use changes resulting from the new Central Corridor Transit Line.

Timeline and Final Product: *Healthy Corridor for All* will begin in Summer 2010. Preliminary findings are anticipated in late Winter of 2011 with a final report available in Spring of 2011, with monitoring continuing into Fall 2011. The final product will be a report detailing the findings of the HIA and recommendations. The conveners and the Community Steering Committee will hold a community meeting to share preliminary findings and recommendations with stakeholders in late Winter or early Spring of 2011.

HIA Goals:

- Assess the impacts of the CCLRT zoning on overall community health, health inequities by race, income, and place, and underlying conditions that determine health in the Central Corridor and the East side.
- Ensure positive health benefits are maximized and negative health impacts are addressed by the decision-making process.
- Empower Central Corridor and East side local communities to effectively and meaningfully engage in the CCLRT zoning process.

HIA Core Values:
The core values that will guide this HIA include:

- Equity
- Racial justice
- Community empowerment
- Collaboration
- Accountability
- Scientific integrity

HIA Collaborators:
These goals and values are central to how the *Healthy Corridor for All* project team will carry out the HIA, from the scoping to the monitoring phase. To support this process, the project team will establish a voluntary Community Steering Committee made up of community stakeholders and a Technical Advisory Panel composed of partners with technical expertise relevant to the project.

The *Community Steering Committee*, made up of constituency-based organizations representing or serving community members, particularly low-income people and people of color, living and working in the East and Central Corridor communities, will be at the center of the HIA. Specifically, the Community Steering Committee will make key HIA decisions and help drive HIA activities, including developing a scope, identifying indicators, developing recommendations, and communicating findings. There may also be opportunities to collect information. Community Steering Committee members have a commitment to improving community health and well-being, promoting equity and have a commitment to grass-roots community building in East and Central Corridor communities

The *Technical Advisory Panel*, made up of agencies, organizations and individuals with an interest in the rezoning process, will provide technical expertise to the HIA. The Committee will review the scoping plan, review assessment methods and findings, share qualitative and quantitative data, and participate in the monitoring process. Technical Advisory Panel members have a commitment to improving health and well-being in the City of St. Paul and the region. Members will commit to providing technical knowledge, data and resources to the Community Steering Committee to conduct the HIA.

The Community Steering Committee will convene for up to five in-person meetings of 3-5 hours each over a one year (ending in July 2011). Technical Advisory Panel members will be consulted as needed. Specific roles and responsibilities of both groups are described in detail below.

Principles of Collaboration:
Membership in both groups is strictly voluntary. However, a number of agreements and commitments must be made in order to participate on either the Steering Committee or the Technical Advisory Panel. Participants must agree to the delineated HIA goals (see above), and also to the core values of equity, racial justice, community empowerment, collaboration, accountability, and scientific integrity. This means that members will work together to conduct the HIA in accordance with these goals and values, and will not challenge these goals and values throughout the process.

In addition, ground rules for participation include:

- Providing constructive and proactive input (rather than obstructive and reactive input)
- Practicing solution-seeking "both/and" thinking (rather than "either/or" thinking)

- Being inclusive by respecting different priorities and concerns
- Trying on new ideas and perspectives
- Attending t all Community Steering Committee meetings, or finding a proxy in the case of unavoidable absence
- Providing feedback and reviewing HIA materials as requested
- Being responsive to outreach regarding the needs of the HIA

Decision-making Process:

During the health impact assessment, the Community Steering Committee and the conveners will be making decisions on the direction of the project. The Technical Advisory Panel will be providing technical guidance and support. The Community Steering Committee will be making many decisions, including: what issues to include in the HIA and how to define the groups of special concern; what information is used in the findings; what the recommendations are; and how to use the HIA to take action.

During the HIA process, decisions should be made by consensus whenever possible. Participants will attempt to bring issues to each other's attention to avoid making unilateral decisions. The partners will recognize and consider different perspectives. The conveners and the Technical Advisory Panel will work with the Community Steering Committee to ensure empirical integrity.

Leadership Team

ISAIAH, TakeAction Minnesota and PolicyLink will manage and conduct the HIA and convene the Community Steering Committee and Technical Advisory Panel. ISAIAH is the primary grantee and fiscal agent for the project. TakeAction Minnesota's Hmong Organizing Program is ISAIAH's primary local partner and PolicyLink is the project's Technical Partner. These organizations work closely together, each with different, but complementary roles:

- ISAIAH and Take Action Minnesota
  - *Role:* ISAIAH is fiscal agent and lead organization for the project. ISAIAH leads the process to identify and invite key constituencies into the *Healthy Corridor for All* Community Steering Committee and Technical Advisory Panel. ISAIAH also leads the coordination between the local partners and Technical Partner. TakeAction Minnesota's Hmong Organizing Program is particularly focused on inviting the participation of Hmong leaders in the project, and assists with general community engagement. Both organizations collaborate to understand the political dynamics relevant to the success of the HIA and the timelines, processes, and materials relevant to the HIA analysis.
  - *HIA Project Manager:* ISAIAH has hired, on contract, a project manager for the HIA. The project manager will manage the stakeholder engagement, and serve as the point-person for the project, acting as the primary contact, coordinator and communicator between the various groups and agencies cooperating on the project.
- PolicyLink
  - *Role:* Technical partner directing the research and empirical HIA process and supporting the overall HIA. PolicyLink will lead the technical aspects of the HIA with support from partners

Stakeholder Engagement:

Stakeholder:

H-3

Definition: Stakeholders are persons or groups who are directly or indirectly affected by a project, as well as those who may have interests in a project or the ability to influence its outcome, either positively or negatively.

Engagement: A set of high-level stakeholders will engage in the HIA through participating on a Community Steering Committee. The Steering Committee will engage a broader group of stakeholders through individual conversations, community meetings, interviews, surveys and focus groups. The conveners of the HIA will also hold meetings with stakeholders, hold community meetings to share HIA findings with interested stakeholders, and strive to engage with as many stakeholders as possible.

Community Steering Committee:

Definition: A group of high level stakeholders who are responsible for providing guidance on overall direction of the HIA. They will help to obtain strategic input and buy-in from a larger set of stakeholders.

- Purpose
    - o Provide strategic direction for the scope and implementation of the HIA representing the views of the group or organization the Committee member represents
    - o Review and provide input on data, materials, and analyses developed throughout the HIA
    - o Develop recommendations based on HIA analysis
    - o Support HIA to ensure partnership and linkage to other stakeholders and key relevant processes
    - o Mobilize and sustain high level of engagement, political commitment and momentum to achieve the HIA objectives
    - o Identify available resources and activities relevant to the HIA
    - o Provide a communication channel to other stakeholders not formally represented on the Committee
    - o Monitor ongoing HIA progress
- Internal Process
    - o Potential Community Steering Committee members will be identified through ISAIAH, TakeAction Minnesota and the Community Steering Committee based on representation of key constituent groups located in the areas affected by Light Rail.
    - o Membership is organizational and not individual.
    - o Any members joining the Steering Committee will sign onto an understanding of the goals and purposes of the HIA, and will ensure their participation is constructive to that end.
    - o New members or alternates will accept all decisions, analyses and input provided in the past in order to engage in present and future activities.
    - o Committee decisions will be made on a consensus basis. Dissenting views may be recorded if required.
    - o Committee members may appoint alternates to replace the representatives in the event of an absence.
    - o Committee members' input and decisions must be received by the deadlines requested. The HIA efforts and activities will move forward based on input and decisions received by indicated deadlines.
    - o The Community Steering Committee will convene in-person a minimum of five times for the following purposes:

H-4

1. Launch of Community Steering Committee and Development of HIA Scope (July 12th and 13th from 6pm to 9pm).
2. Input on baseline data analysis and next steps (date to be determined).
3. Input on predictions on health impact of proposal (date to be determined).
4. Development and prioritization of recommendations based on analysis (date to be determined).
5. Sharing full analysis with larger stakeholder group (date to be determined).
   - In addition to scheduled meetings, the Committee may hold periodic conference calls with a set agenda on an as-needed basis.
   - Individual members may be called upon for expertise on issues within their area of knowledge.

Meeting notes will be prepared by the chairs of the Committee with a record of what decisions were made and what actions need to be taken and by whom. The minutes will be prepared within two weeks of the meeting and sent to all members via email. Committee members will have the opportunity to add to the notes if anything is left out.

Technical Advisory Panel:
Definition: A group of researchers, experts, and government staff with expertise on key issues related to the Central Light Rail Corridor HIA project.

- Purpose
  - Using best possible technical expertise, review and improve HIA methodologies and analyses
  - Identify and, when possible, provide information, data, activities and resources
  - Support HIA to ensure linkages to other technical advisors and key relevant processes
  - Share perspective of organization or agency Advisor is representing
  - Help identify appropriate monitoring mechanisms to examine the progress of several health indicators
- Internal Process
  - Potential Technical Advisory Panel members will be identified through ISAIAH, TakeAction Minnesota and PolicyLink and Community Steering Committee based on a review of the Committee's technical needs, data needs, and an overall understanding of the scientific context of the Central Corridor Light Rail HIA.
  - Membership is organizational and not individual.
  - Technical Advisory Panel members may appoint alternates to replace them as representatives in the event of an absence.
  - Group members' input and decisions must be received by the deadlines requested. The HIA efforts and activities will move forward based on input and decisions received by indicated deadlines.
  - New members or alternatives will accept all decisions, analyses and input provided in the past in order to engage in present and future activities.
  - The Advisory Group will not be formal voting members of the Community Steering Committee but will provide guidance, advice and materials to the Committee advice as needed, in order to help the Committee in their decision-making.
  - The Advisory Panel members will join as many of the Community Steering Committee meetings as possible to provide technical advice. Their expertise during the meetings will be valuable. The foreseeable Committee meetings include approximately five in-person meetings between July 2010 and May 2011. These meetings include the following:

1. Launch of Steering Committee and Development of HIA Scope (July 12th and 13th from 6pm to 9pm).
2. Input on baseline data analysis and next steps (date to be determined).
3. Input on predictions on health impact of proposal (date to be determined).
4. Development and prioritization of recommendations based on analysis (date to be determined).
5. Sharing full analysis with larger stakeholder group (date to be determined)
- Individual members may be called upon for expertise on issues within their area of knowledge.

Appendix I –Opportunities for Stakeholder Involvement in Each Step of HIA

The following is an excerpt from the *Guidance and Best Practices for Stakeholder Participation in Health Impact Assessments* (Stakeholder Participation Working Group 2010) and summarizes the opportunities for stakeholder engagement in HIA, as outlined by the North American HIA Practice Standards Working Group (2010).

HIA Standards for Practitioners	Section
Process Oversight: *Intended to be used throughout all the stages of the HIA*	
Essential ▪ Accept and utilize diverse stakeholder input.	1.5
Recommended ▪ Have a specific engagement and participation approach that utilizes available participatory or *deliberative methods* suitable to the needs of stakeholders and context	1.6
Screening Stage: *Deciding whether an HIA is needed, feasible, and relevant*	
Essential ▪ Understand stakeholder concerns in order to determine potential health effects.	2.2.3
▪ Identify and notify stakeholders of decision to conduct a HIA.	2.3
Recommended ▪ Identify stakeholders to potentially partner with a HIA.	
▪ Seek diverse stakeholder participation in screening the target policy or HIA plan.	
Scoping Stage: *Deciding which health impacts to evaluate and evaluation methodology*	
Essential ▪ Use input from multiple perspectives to inform *pathways* (between the policy, plan or project and key health outcomes). Use multiple avenues to solicit input (from stakeholders, affected communities, decision makers).	3.1
▪ Ensure a mechanism to incorporate new feedback from stakeholders on the scope of the HIA.	3.7
Recommended ▪ Work with diverse stakeholders to prioritize key elements of analysis.	
▪ Seek feedback from stakeholders on HIA scope.	
Assessment Stage: *Using data, research, and analysis to determine the magnitude and direction of potential health impacts*	
Essential ▪ Use local knowledge as part of the evidence base.	4.2.1
▪ Work to engage all stakeholders in data collection.	4.2.4
Recommended ▪ Seek feedback from stakeholders on draft findings.	
Recommendations: *Providing recommendations to manage the identified health impacts and improve health conditions*	
Essential ▪ Use expert guidance to ensure recommendations reflect effective practices.	5.2
Recommended ▪ Work with community and other stakeholders to identify and prioritize recommendations.	
▪ Seek input on recommendations.	
Reporting & Communication: *Sharing the results, recommendations*	
Essential ▪ Summarize primary findings and recommendations to allow for stakeholder understanding, evaluation, and response.	6.2
▪ Document stakeholder participation in the full report.	6.3
▪ Make an inclusive accounting of stakeholder values when determining recommendations.	6.5
▪ Allow for, and formally respond to, critical review from stakeholders, and make the report publicly accessible.	6.6-7
Recommended ▪ Seek diverse input on draft final report.	
▪ Work with stakeholders to build their capacity to understand and articulate the findings of the HIA.	
Monitoring: *Tracking how the HIA affects the decision and its outcomes*	
Essential ▪ Plan should address reporting outcomes to decision makers.	7.2
▪ Monitoring methods and results should be made available to the public.	7.4
Recommended ▪ Involve interested stakeholders in monitoring outcomes.	

Appendix J –Risk Assessment Technique for Impact Prioritization

The following is an excerpt from the Point Thomson Project HIA:

2.2.1.1   Risk Assessment Matrix

While there are numerical risk-based environmental standards that regulate biota, air, water and s oil, there are no similar quantitative regulatory endpoints for public-health outcomes. Winkler 2010 proposes a risk assessment technique that ranks the significance of identified health impacts allowing health planners prioritize management actions. The entire rating is based on a modified Delphi approach (Rowe and Wright, 1999), a technique used in judgment and forecasting situations where pure model-based statistical methods are not practicable.

The HIA team performed this evaluation, as fully described in Winkler 2010 by drawing on

(i)   Available health baseline data from the literature review;
(ii)   Review of the project context, alternatives and developments;
(iii)   Review of pertinent sections of the Point Thomson Project Environmental Impact Statement, particularly the Socioeconomic, Environmental Justice, Subsistence, and Transportation section; and
(iv)   Information and recommendations generated by a panel of Alaskan medical and public health professionals.

The HIA team created a worksheet for each of the eight HECs and each of the five alternatives. Each of the 40 worksheets was divided into the project phases: construction, drilling, and operation. The health impact parameters consider:

- **Duration** – determines how long each phase will last; ranked from under a month to beyond the life of the project

- **Magnitude** – evaluates the intensity of the impact, particularly in light of existing baseline conditions

- **Extent** – identifies the localities where the projected impact will be experienced, *e.g.*, local or regional

- **Likelihood** – evaluates the probability that the impact will occur

- **Nature** – determines whether the impact is direct, indirect or cumulative

- **Impact** – evaluates whether the impact is positive or negative, *i.e.*, whether the impact will promote or progress, degrade or detract from the well-being of defined communities or populations

- **Scoring** – as described in Figure 27 and Figure 28 below.

For the risk analysis, a 4 -step procedure was developed that is illustrated on t he risk assessment matrix (Figure 27 and 28), as modified from Winkler 2010, and as presented below.

Figure 27 Step 1 of 4-Step Risk Assessment Matrix

Step 1				
	Consequences			
Impact Level (score)	**A – Health Effect**	**B- Duration**	**C-Magnitude**	**D- Extent**
Low (0)	Effect is not perceptible	Less than 1 month	Minor intensity	Local/Project Area
Medium (1)	Effect results in annoyance, minor injuries or illnesses that do not require intervention	Short-term: 1-12 months	Those impacted will be able to adapt to the impact with ease and maintain pre-impact level of health	Local/Zone 1: Kaktovik and Nuiqsut
High (2)	Effect resulting in moderate injury or illness that may require intervention	Medium-term: 1 to 6 years	Those impacted will be able to adapt to the health impact with some difficulty and will maintain pre-impact level of health with support	Zone 2: Prudhoe Bay/Deadhorse AP Barrow
Very high (3)	Effect resulting in loss of life, severe injuries or chronic illness that requires intervention	Long-term: more than 6 years/life of project and beyond	Those impacted will not be able to adapt to the health impact or to maintain pre-impact level of health	Rest of Alaska US Global

In Step 1, the extent of the four different consequences — (A) effect; (B) duration; (C) magnitude; and (D) extent—is rated according to the criteria set forth in Figure 28. The output of this rating is a score between 0 and 3 for each consequence, depending on the estimated impact level:

- Low (score = 0)
- Medium (score=1)
- High (score=2)
- Very high (score=3).

Figure 28 Steps 2, 3, and 4 of 4-Step Risk Assessment Matrix

Step 2	Step 3						
Severity Rating	**Likelihood Rating**						
(Magnitude + Duration + Geographic Extent + Health Effect)	Extremely Unlikely < 1%	Very Unlikely 1-10%	Unlikely 10-33%	About as Likely as Not 33-66%	Likely 66-90%	Very Likely 90-99%	Virtually Certain > 99%
Low (0-3)	♦	♦	♦	♦	♦♦	♦♦	♦♦
Medium (4-6)	♦	♦	♦	♦♦	♦♦	♦♦	♦♦♦
High (7-9)	♦♦	♦♦	♦♦	♦♦♦	♦♦♦	♦♦♦	♦♦♦♦
Very high (10-12)	♦♦♦	♦♦♦	♦♦♦	♦♦♦♦	♦♦♦♦	♦♦♦♦	♦♦♦♦
Step 4	Impact Rating						
Key: Low ♦ Medium ♦♦ High ♦♦♦ Very High ♦♦♦♦							

In Step 2, as shown in Figure 28, the scores of the consequences are summed up and based on the value the impact severity is assigned as follows:

- Low (0–3)
- Medium (4–6)
- High (7–9)
- Very high (10–12).

In Step 3 the likelihood of the impact to occur is assessed according to the following definitions, as presented in IPCC 2007:

- Exceptionally unlikely   < 1   percent probability
- Very unlikely   1-10 percent probability
- Unlikely   10-33 percent probability
- About as likely as not: 33-66 percent probability
- Likely: 66-90 percent probability
- Very likely: 90-99 percent probability
- Virtually certain: > 99 percent probability.

Step 4 entails the final significance rating, which is identified through the intersection of the impact severity and the likelihood of the impact to occur, as shown in Figure 28.

A low significance indicates that the potential health impact is one where a negative effect may occur from the proposed activity; however, the impact magnitude is sufficiently small (with or without mitigation) and well within accepted levels, and/or the receptor has low sensitivity to the effect.

Impacts classified with a medium significance and above require action so that predicted negative health effects can be mitigated to as low as reasonably practicable (Winkler 2010). An impact with

J-3

high or very high significance will affect the proposed activity, and without mitigation, may present an unacceptable risk. The significance is simply stated as positive (e.g. improvement of health services). If there is a negative accentuation of the health impact compared to the baseline condition, this is indicated in the risk assessment matrix by the use of a + sign to indicate a positive impact or a − sign to indicate a negative impact.

www.ingramcontent.com/pod-product-compliance
Lightning Source LLC
Chambersburg PA
CBHW081439170526

45166CB00008B/2247